ibvt-Schriftenreihe

Schriftenreihe des Institutes für Bioverfahrenstechnik
der Technischen Universität Braunschweig

Herausgegeben von Prof. Dr. Rainer Krull

Band 81

I0131044

Cuvillier-Verlag
Göttingen, Deutschland

Herausgeber
Prof. Dr. Rainer Krull
Institut für Bioverfahrenstechnik
TU Braunschweig
Rebenring 56, 38106 Braunschweig
www.ibvt.de

Hinweis: Obgleich alle Anstrengungen unternommen wurden, um richtige und aktuelle Angaben in diesem Werk zum Ausdruck zu bringen, übernehmen weder der Herausgeber, noch der Autor oder andere an der Arbeit beteiligten Personen eine Verantwortung für fehlerhafte Angaben oder deren Folgen. Eventuelle Berichtigungen können erst in der nächsten Auflage berücksichtigt werden.

Bibliographische Informationen der Deutschen Nationalbibliothek
Die Deutsche Nationalbibliothek verzeichnet diese Publikation in der Deutschen Nationalbibliographie; detaillierte bibliographische Daten sind im Internet über *http://dnb.d-nb.de* abrufbar.
1. Aufl. – Göttingen: Cuvillier, 2019

© Cuvillier-Verlag · Göttingen 2019
    Nonnenstieg 8, 37075 Göttingen
    Telefon: 0551-54724-0
    Telefax: 0551-54724-21
    www.cuvillier.de

1. Auflage, 2019
Gedruckt auf säurefreiem Papier

    ISBN 978-3-7369-7025-0
    eISBN 978-3-7369-6025-1
    ISSN 1431-7230

Bioengineering at the micro-scale:

Design, characterization and validation of

microbioreactors

Bei der Fakultät für Maschinenbau
der Technischen Universität Carolo-Wilhelmina zu Braunschweig

zur Erlangung der Würde
einer Doktor-Ingenieurin (Dr.-Ing.)
eingereichte Dissertation

von

Ing. Química, M. Sc. Susanna Maria Lladó Maldonado
aus Barcelona

eingereicht am: 07.06.2018

mündliche Prüfung am: 27.09.2018

Vorsitz: Prof. Dr. Andreas Dietzel

Gutachter: Prof. Dr. Rainer Krull

Prof. Dr. Ulrich Krühne

2019

I am among those who think that science has great beauty.

Marie Skłodowska-Curie

Acknowledgements

The present thesis was developed during my work as PhD candidate at the Institute of Biochemical Engineering of the Technische Universität Braunschweig and it was funded by the People Programme (Marie Curie Actions, Multi-ITN) of the European Union's Seventh Framework Programme for research, technological development and demonstration within the project *EUROMBR — European network for innovative microbioreactor applications in bioprocess development* (Project ID 608104). I also received an additional six-month funding from the *Promotionsabschlussförderung* program of the Gleichstellungsbüro of the TU Braunschweig.

First of all, I would like to thank my supervisor Prof. Dr. habil. Rainer Krull, my *Doctorfather*, for his passion, engagement, and all the support all the way through the project.

I would like to show my gratitude to Prof. Dr. Ulrich Krühne, Process and Systems Engineering Center (PROSYS), Department of Chemical and Biochemical Engineering, Technical University of Denmark, for his invaluable help with *CFD*, and for accepting being part of the dissertation committee.

Thanks to Prof. Dr. Andreas Dietzel from the Institute of Microtechnology, TU Braunschweig, for taking over the presidency of the dissertation committee.

A huge thank to Detlev Rasch for being extremely helpful in the laboratory set-up. Thanks to Dr.-Ing. Astrid Edlich for reviewing my work.

I would like to thank the *EUROMBR* family for all the interesting discussions and other unforgettable moments. *EUROMBR* has given me access to a great network of international experts in the field, the chance of participating into high quality trainings and conferences, and living new enriching experiences. A special thanks to the cooperation partners for the fruitful collaborations: Prof. Dr. Ulrich Krühne (Process and Systems Engineering Center (PROSYS), Department of Chemical and Biochemical Engineering, Technical University of Denmark), Prof. Dr. Torsten Mayr and Shiwen Sun (Institute of Analytical Chemistry and Food Chemistry, Graz University of Technology, Austria), Dr. Adama M. Sesay and Peter Panjan

(Measurement Technology Unit, University of Oulu, Kajaani, Finland), Dr. Juan M. Bolivar and Donya Valikhani (Institute of Biotechnology and Biochemical Engineering, Graz University of Technology, Austria) and Prof. Nicolas Szita and Dr. Marco Marques (Department of Biochemical Engineering, University College London, United Kingdom).

I am thankful to the most dedicated and motivated students Jana Krull (in cooperation with the Microfluidics Laboratory in the Department of Biochemical Engineering, University College London, United Kingdom) and Leopold Heydorn (in cooperation with the Institute of Biotechnology and Biochemical Engineering, Graz University of Technology, Austria) for pursuing their master thesis in the field of microbiroeactors.

My greatest gratitude to the *ibvt* colleagues for the pleasant working atmosphere, and specially my officemates Christina, Lasse, and Leopold who would listen and help in any moment.

I would also thank to my flatmates from the Bültenweg WG, for being my family away from home. I would especially thank my dearest friend Elena, for taking care of me, being always in the positive side, and laughing with me.

To my family and friends in Spain, that despite the distance I could feel them very close to me.

Thank you all for accompanying me during the dark and cold days but also the sunny days.

Susanna Maria Lladó Maldonado
Braunschweig, May 2019

Publications of the thesis

This thesis was performed during my doctoral research at Institut für Bioverfahrenstechnik (*ibvt*), Technische Universität Braunschweig, as part of the *EUROMBR* project - *European network for innovative microbioreactor applications in bioprocess development* (Project ID 608104) funded by the People Programme (Marie Curie Actions, Multi-ITN) of the European Union's Seventh Framework Programme for research, technological development and demonstration.

Aspects of this thesis have been published previously:

Chapter 2:
Krull R, Lladó Maldonado S, Lorenz T, Büttgenbach S, Demming S. 2016. Microbioreactors. In: Dietzel A, editor. *Microsystems Pharmatechnology. Manipulation of fluids, particles, droplets, and cells.* Chapter 4, Series Microsystems and Nanosystems, ISBN: 978-3-319-26918-4 (Print) 978-3-319-26920-7 (Online). Cham Heidelberg New York Dordrecht London: Springer International Publishing, pp. 99–152.

Chapter 4:
Lladó Maldonado S, Rasch D, Kasjanow A, Bouwes D, Krühne U, Krull R. 2018. Multiphase microreactors with intensification of oxygen mass transfer rate and mixing performance for bioprocess development. *Biochem. Eng. J.* 139:57–67.

Chapter 5:
Lladó Maldonado S, Panjan P, Sun S, Rasch D, Sesay AM, Mayr T, Krull R. 2019a. A fully online sensor-equipped, disposable multiphase microbioreactor as a screening platform for biotechnological applications. *Biotechnol. Bioeng.* 116:65–75.

Chapter 6:
Lladó Maldonado S, Krull J, Rasch D, Panjan P, Sesay AM, Marques MPC, Szita N, Krull R. 2019b. Application of a multiphase microreactor chemostat for the determination of reaction kinetics of *Staphylococcus carnosus*. *Bioprocess Biosyst. Eng.* 42:953–961.

Articles in Journals

Peterat G, Lladó Maldonado S, Edlich A, Rasch D, Dietzel A, Krull R. 2015. Bioreaktionstechnik in mikrofluidischen Reaktoren. Chemie Ing. Tech. 87:505–517.

Lladó Maldonado S, Rasch D, Kasjanow A, Bouwes D, Krull R. 2016. Hydrodynamic and mass transfer characteristics of a microbubble column-bioreactor. Chemical and Biological Microsystems Society, Dublin, pp 1483–1484.

Lladó Maldonado S, Rasch D, Kasjanow A, Bouwes D, Krühne U, Krull R. 2018. Multiphase microreactors with intensification of oxygen mass transfer rate and mixing performance for bioprocess development. *Biochem. Eng. J.* 139:57–67.

Lladó Maldonado S, Panjan P, Sun S, Rasch D, Sesay AM, Mayr T, Krull R. 2019a. A fully online sensor-equipped, disposable multiphase microbioreactor as a screening platform for biotechnological applications. *Biotechnol. Bioeng.* 116:65–75.

Lladó Maldonado S, Krull J, Rasch D, Panjan P, Sesay AM, Marques MPC, Szita N, Krull R. 2019b. Application of a multiphase microreactor chemostat for the determination of reaction kinetics of *Staphylococcus carnosus*. *Bioprocess Biosyst. Eng.* 42:953–961.

Chapter Books

Krull R, Lladó Maldonado S, Lorenz T, Büttgenbach S, Demming S. 2016. Microbioreactors. In: Dietzel A, editor. *Microsystems Pharmatechnology. Manipulation of fluids, particles, droplets, and cells.* Chapter 4, Series Microsystems and Nanosystems, ISBN: 978-3-319-26918-4 (Print) 978-3-319-26920-7 (Online). Cham Heidelberg New York Dordrecht London: Springer International Publishing, pp. 99–152.

Oral Presentations at Congresses

Krull R, Lladó Maldonado S, Peterat G. 2015. Reaktionskinetische Untersuchungen zur Chemostatkultivierung von *Saccharomyces cerevisiae* mit einem Mehrphasen-Mikroblasensäulenreaktor. Annual colloquium 2015 DFG SPP 1740 *Reactive bubble flow, Honorary colloqium in memoriam Prof. Dr.-Ing. Norbert Räbiger*, TU Hamburg-Harburg, Harburg, August 2015.

Lladó Maldonado S, Krull J, Rasch D, Kasjanow A, Bouwes D, Krühne U, Krull R. 2017. Multiphase microreactors with intensification of oxygen mass transfer rate and mixing performance for bioprocess development, 4th International Conference Implementation of Microreactor Technology in Biotechnology (IMTB 2017), Book of extended abstracts: Session B: Cells within microreactors, BO4, 48 – 49, Bled, Slovenia, April 2017.

Frey LJ, Rasch D, Lladó Maldonado S, Meinen S, Dietzel A, Krull R. 2017. Development of a microbiorector system for biopharmaceutical applications and analysis of scale down effects. 4th International Conference Implementation of Microreactor Technology in Biotechnology (IMTB 2017), Book of extended abstracts: Session D: Process intensification and integration, DO3, 89 – 90, Bled, Slovenia, April 2017.

Lladó Maldonado S, Rasch D, Frey LJ, Meinen S, Dietzel A, Krull R. 2017. Cultivation at the microscale - development of multiphase microreactors as screening platform for biotechnological and pharmaceutical applications", 3rd International Symposium on Multiscale Multiphase Process Engineering (MMPE), Proc. L-16, 110-111, Toyama, Japan, May 2017.

Lladó Maldonado S, Rasch D, Kasjanow A, Bouwes D, Krühne U, Krull R. 2017. Bioengineering at the microscale: Process characterization of a microbubble column-bioreactor, 2nd Braunschweig International Symposium on Pharmaceutical Engineering Research – SphERe, Center of Pharmaceutical Engineering, Braunschweig, Germany, September 2017.

Lladó Maldonado S, Krull J, Heydorn RL, Valikhani D, Rasch D, Nidetzky B, Bolivar JM, Krull R. 2017. Cultivation and biotransformation at the microscale: intensification of oxygen mass transfer and mixing in multiphase microreactors, 10th World Congress of Chemical Engineering (WCCE 10) and 4th European Congress of Applied Biotechnology (ECAB4), Barcelona, Spain, October 2017.

Poster Presentations at Congresses

Lladó Maldonado S, Rasch D, Krull R. 2015. Implementation of a CO_2 sensor in microbioreactors, 3rd International Conference Implementation of Microreactor Technology in Biotechnology (IMTB 2015), Book of extended abstracts: Session B: Cells within microstructed devices, BP2, 75 – 76, Opatja, Croatia, May 2015.

Lladó Maldonado S, Sun S, Rasch D, Mayr T, Krull R. 2016. Implementation of optical chemical sensors in a microbubble column-bioreactor, EUROPT(R)ODE XIII, Graz, Austria, March 2016.

Lladó Maldonado S, Rasch D, Kasjanow A, Bouwes D, Krull R. 2016. Hydrodynamic and mass transfer characteristics of a microbubble column-bioreactor, 20th International Conference on Miniaturized Systems for Chemistry and Life Sciences (MicroTAS), Poster M209i, Dublin, Ireland, October 2016.

Lladó Maldonado S, Rasch D, Krull R. 2017. Cultivation at the microscale: Characterization of mixing performance and oxygen transfer in a microbubble column-bioreactor, Spring meeting of the Deutsche Technion-Gesellschaft, Leibnizhaus, Hannover, Hannover, Germany, March 2017.

Frey L, Rasch D, Lladó Maldonado S, Meinen S, Dietzel A, Krull R. 2017. A novel microscale cell culturing system using electrodynamic mixing for biopharmaceutical application, 3rd International Symposium on Multiscale Multiphase Process Engineering (MMPE), Proc. P-15, 265-266, Toyama, Japan, May 2017.

Heydorn RL, Lladó Maldonado S, Panjan P, Bolivar JM, Valikhani D, Sesay AM, Nidetzky B, Krull R. 2017. Bioconversion at microscale: Application of oxygen dependent enzyme immobilizates in microfluidised bed reactors, 2nd Braunschweig International Symposium on Pharmaceutical Engineering Research – SphERe, Center of Pharmaceutical Engineering, Braunschweig, Germany, September 2017.

Krull J, Lladó Maldonado S, Rasch D, Marques MPC, Szita N, Krull R. 2017. Cultivation of *Staphylococcus carnosus* at micro-scale – Influence of different reactor performances, 2nd Braunschweig International Symposium on Pharmaceutical Engineering Research – SphERe, Center of Pharmaceutical Engineering, Braunschweig, Germany, September 2017.

Ostsieker H, Frey L, Vorländer D, Müller B, Lladó Maldonado S, Rasch D, Mayr T, Krull R. 2018. Innovative microbioreactor applications in bioprocess development, 9th Workshop of Chemical and Biological Micro Laboratory Technology, Elgersburg, Ilmenau, Germany, February 2018.

Presentations at Project Meetings of the *EUROMBR* Consortium

Lladó Maldonado S, Rasch D, Krull R. 2015. Bioengineering at the micro-scale: Enhanced mixing on small scales and cultivation in microfluidic screening bioreactors, EUROMBR Meeting 2, University College of London, United Kingdom, February 2015.

Lladó Maldonado S, Rasch D, Krull R. 2015. Real-time online monitoring of an aerobic batch cultivation in a microbubble column, EUROMBR Meeting 3 and 5, TU Graz, Austria, September 2015.

Lladó Maldonado S, Rasch D, Kasjanow A, Krull R. 2016. Characterization of a highly sensor integrated microbubble column-bioreactor, EUROMBR Meeting 6, TU Braunschweig, Germany, May 2016.

Lladó Maldonado S, Rasch D, Krull R. 2016. Evaluation of the mixing performance in a microfluidic reactor, EUROMBR Meeting 7, Technical University of Denmark, Denmark, September 2016.

Lladó Maldonado S, Rasch D, Krull R. 2017. Cultivation at the microscale – Evolution of a microbubble column-bioreactor for bioprocess development. EUROMBR Meeting 8, Magna Græcia University of Catanzaro, Italy, October 2017.

Student's Theses

Krull J. 2017. Cultivation of *Staphylococcus carnosus* at micro-scale - Influence of different reactor performances, Master thesis (58) Institute of Biochemical Engineering, TU Braunschweig/ Microfluidics Laboratory in Biochemical Engineering, University College London (London, United Kingdom).

Heydorn RL. 2017. Intensified bioconversion at microscale: Application of oxygen dependent enzyme immobilizates in microfluidised bed reactors, Master thesis (59), Institute of Biochemical Engineering, TU Braunschweig/ Institute of Biotechnology and Biochemical Engineering, TU Graz (Graz, Austria).

Abbreviations and Symbols

Abbreviations

BF	borofloat
cµBC	cuvette-based microbubble column-bioreactor
CAD	computer-aided design
CFD	computational fluid dynamics
fps	frames per second
gµBC	glass-based microbubble column-bioreactor
GOx	glucose oxidase
hMBR	horizontal microbioreactor
HF	hydrofluoric acid
HTP	high-throughput processing
HTS	high-throughput screening
I	light intensity
LoC	lab-on-a-chip
MBR	microbioreactor
mSTR	miniaturised stirred tank reactor
MTP	microtiter plate
PB	Prussian blue
PBS	phosphate buffered saline
PDMS	polydimethylsiloxane
PIV	Particle Image Velocimetry
PMMA	polymethyl methacrylate
PtTPTBPF	platinum(II) meso-tetra(4-fluorophenyl)tetrabenzoporphyrin
RGB	Red Green Blue colour model
RMS	Root Mean Square
TAS	total chemical analysis system
THF	tetrahydrofuran
U	unit of enzyme activity (µmol/min)
µBC	microbubble column-bioreactor
µFBR	microfluidised bed reactor

Nomenclature

a	interfacial area per unit volume (m^2/m^3)
A	cross-sectional area (m^2)
A_{obs}	observed activity (U/L)
c	concentration (mol/L)
c^*	saturation concentration (mol/L)
c_L	oxygen concentration in the liquid phase (mmol/L)
c_S	substrate concentration (g/L)
c_{CDW}	biomass concentration (g_{CDW}/L)
CDW	cell dry weight (g)
d	diameter (m)
d_{vs}	Sauter mean bubble diameter (m)
D	dilution rate (1/h)
DO	dissolved oxygen (%)
$dphi$	phase shift (°)
F	flow rate (L/s)
G^*	normalized green channel data
H	height of the fluid in the column (m)
k_L	liquid-phase mass transfer coefficient (m/h)
$k_L a$	volumetric liquid-phase mass transfer coefficient (1/h)
K_S	Monod constant (g/L)
m_S	maintenance coefficient ($g_S/(g_{CDW} \cdot h)$)
N	number of bubbles
OD	optical density (-)
OTR	volumetric oxygen transfer capacity (mmol/(L·h))
OUR	volumetric oxygen uptake rate (mmol/(L·h))
p	pressure (bar)
Pr	biomass-related productivity (g_{CDW}/(L·h))
P/V	volumetric power input (W/m^3)
q_S	specific substrate uptake rate ($g_S/(g_{CDW} \cdot h)$)
Q	volumetric flow rate (m^3/s)
Re	Reynolds number (-)
S	surface (m^2)

sOUR	specific oxygen uptake rate (mmol/($g_{CDW} \cdot h$))
t	time (s)
u	superficial velocity (m/s)
υ	velocity (m/s)
V	volume (L)
x, y, z	Cartesian coordinate (m)
$Y_{X/S}$	substrate-related biomass yield coefficient (g_{CDW}/g_S)

ε	hold-up
θ	mixing time (s)
μ	liquid dynamic viscosity ($N \cdot s/m^2$)
μ	specific growth rate (1/h)
η	effectiveness factor (%)
σ	surface tension between liquid and gas (N/m)
ρ	liquid density (kg/m^3)

Indices

b	bubble
G	gas
h	hydraulic
L	liquid
tot	total

Abstract

The miniaturisation of bioreactors to the microliter-scale and the integration of online sensors for monitoring the most important process variables during the cultivations is a promising approach for the screening and optimization of cultivation and biocatalysis processes, enabling information-rich, parallelised and cost-effective experiments under well-controlled environmental conditions. The main advantages of microbioreactors (*MBRs*) are minimization of space and reagents, their easy manipulation, and their high-throughput screening potential, which make them very interesting tools to develop bioprocesses.

This thesis is focused on the development of *MBRs* and the sensor integration for monitoring optical density (*OD*), dissolved oxygen (*DO*), *pH* and glucose as well as their validation for different biotechnological applications.

On the basis of the developed *PDMS*-based *MBR* from the *mikroPART* project (2011-2014 at the TU Braunschweig), a borosilicate glass-based microbubble column-biroeactor (*gμBC*) (working volume of 60 μL, aeration occurs through a nozzle with Ø = 26 μm) was designed and manufactured by wet etching and powder blasting technology. The *gμBC* proved to have good oxygen transfer capacity, reaching k_La values up to 320 1/h and fast mixing times θ_{95} down to 5.5 s when working at a gas superficial velocity u_G of $2.25 \cdot 10^{-3}$ m/s. The mixing performance was simulated using a simplified *CFD* model, and the tracer profile yielded a good qualitative prediction that was comparable to the experimental results, presenting a tolerable deviation of the mixing times. The *gμBC* was validated as a suitable cultivation screening tool with a batch cultivation of *Saccharomyces cerevisiae* with the real-time online monitoring of *OD* and *DO*.

Furthermore, a cuvette-based microbubble column-bioreactor (*cμBC*) made of polystyrene (working volume of 550 μL, aeration occurs through a nozzle with Ø ≤ 100 μm) was developed and manufactured with online sensors for *pH*, *OD*, *DO* and glucose. The *cμBC* showed homogeneous mixing of the cultivation medium with θ_{95} < 1 s, with high k_La up to 775 1/h at u_G of $8.4 \cdot 10^{-3}$ m/s. The applicability of the *cμBC* for aerobic submerged whole-cell cultivation in batch and chemostat mode was demonstrated with the model organisms *S. cerevisiae* and *Staphylococcus carnosus*.

In addition, the use of the *cμBC* for oxygen-dependent (cell-free) biocatalysis was successfully demonstrated with the example of the model enzyme glucose oxidase immobilization on supports to convert glucose via gluconolactone and hydrogen peroxide to gluconic acid in a microfluidic bed bioreactor (*μFBR*).

The characterization of the developed and manufactured *MBRs* in this work as well as the integration of the online sensors for *OD, DO, pH* and glucose are now the basis for the future development of a consolidated and parallelisable *MBR* system for bioprocess development.

Kurzfassung

Die Miniaturisierung von Bioreaktoren in den Mikroliter-Maßstab und ihre Ausstattung mit Online-Sensorik zur Überwachung der wichtigsten Analyten während der Kultivierung stellt eine vielversprechende Technologie für die Entwicklung von (Ganzzell-) Kultivierungs- und (zellfreien) Biokatalyseprozessen dar, die informationsreiche, parallelisierte und kostengünstige Experimente unter gut kontrollierbaren Umgebungsbedingungen ermöglichen. Die entscheidenden Vorteile des Einsatzes von Mikrobioreaktoren (*MBR*) für die Bioprozessentwicklung liegen vor allem in der Kostenreduzierung von Experimenten mit teuren Spezialreagenzien sowie ihrem Potential für das Hochdurchsatz-Screening.

In der vorliegenden Dissertation bestand die Aufgabe in der Entwicklung von *MBR* und der Sensorimplementierung für die wichtigen Prozessparameter Optische Dichte (*OD*), Gelöstsauerstoff (*DO*), *pH*-Wert und Glukose sowie ihrer Validierung für unterschiedliche biotechnologische Anwendungen.

Aufbauend zu den entwickelten *PDMS*-basierten *MBR* aus dem *mikroPART*-Projekt (2011-2014 an der TU Braunschweig) wurde zunächst eine Borosilikatglas-basierte Mikroblasensäule (*gµBC*) (Arbeitsvolumen 60 µL, Düsenbelüftung, Ø= 26 µm) konstruiert und mittels Nassätz- und Pulverstrahltechnik gefertigt. Mit der *gµBC* konnten geringe Mischzeiten (θ_{95}) von 5,5 s bei Gasleerrohrgeschwindigkeiten (u_G) von $2,25\cdot10^{-3}$ m/s realisiert werden. Durch die aktive Begasung wurden k_La-Werte von bis zu 320 1/h erzielt. Zur Abschätzung der Reaktordurchmischung wurden *CFD*-Simulationen durchgeführt, die eine gute qualitative Vorhersage der Fluiddynamik mit vergleichbaren Resultaten zu den experimentell generierten Ergebnissen sowie eine tolerierbare Abweichung der Mischzeiten ergaben. Die Anwendbarkeit der *gµBC* wurde in Batch-Kultivierungen von *Saccharomyces cerevisiae* mit dem Online-Monitoring von *OD* und *DO* validiert.

Weiterhin wurde eine küvettenbasierte Mikroblasensäule (*cµBC*) aus Polystyrol (Arbeitsvolumen 550 µL, Begasungsdüse, Ø ≤ 100 µm) mit Online-Sensoren für *pH*, *OD*, *DO* und Glukose entwickelt und gefertigt. Mit der *cµBC* konnten eine homogene

Durchmischung des Kultivierungsmediums bei θ_{95} < 1 s, sowie ausreichend hohe k_La-Werte bis 775 1/h bei u_G von $8.4 \cdot 10^{-3}$ m/s realisiert werden. Die Anwendbarkeit der $c\mu BC$ für die aerobe, submerse Ganzzellkultivierung im Batch- und Chemostat-Modus mit den Modellorganismen *S. cerevisiae* und *Staphylococcus carnosus* konnte aufzeigt werden. Darüber hinaus wurde der Einsatz der $c\mu BC$ für sauerstoffabhängige (zellfreie) Biokatalyseprozesse am Beispiel der auf einem Träger immobilisiert Glukoseoxidase zur Umsetzung von Glukose über Glukonolacton und Wasserstoffperoxid zu Glukonsäure in einem Mikrofluidbett-Bioreaktor (*µFBR*) erfolgreich dargestellt.

Mit der Entwicklung und Fertigung der in dieser Arbeit charakterisierten *MBR* sowie der Implementierung der Online-Sensorik für *OD, DO, pH* und Glukose sind nunmehr die Grundlagen für eine zukünftige Entwicklung eines vollständigen und parallelisierbaren *MBR*-Systems für die Entwicklung von Kultivierungs- und Biokatalyseprozessen gelegt.

Content

1 Introduction, objectives and thesis outline

Biotechnological process development includes the screening of microorganisms and the optimization of cultivation conditions for high-yield bioproduction. Biological reaction kinetics, growth behaviour and product formation are dependent on several physicochemical parameters, e.g., temperature, *pH* and nutrient availability (Krull et al., 2016). There is a high demand for cost-effective, parallel and multi-parametric automated methods for the high-throughput screening (*HTS*) of bioprocesses (Zanzotto et al., 2004). Microprocess engineered bioreactor systems have precise control of the microenvironment and allow the collection of desired information for biotechnological process development that is relevant to the production scale (Hegab et al., 2013; Marques and Szita, 2017). Miniaturised bioreactors with a working volume below 1000 µL are known as microbioreactors (*MBRs*). *MBRs* owing to their small size enable serial processing and analysis and, furthermore, can achieve massive parallelisation through efficient miniaturised integrated sensors and multiplexing. (Perozziello et al., 2012; Zhang et al., 2007). For cultivations, the simultaneous detection of several analytical parameters is mandatory for comprehensive process information, making online analysis here essential since the small reaction volumes of *MBRs* exclude elaborate sampling.

MBRs have many advantages, but it is important to keep in mind that they could also present some drawbacks related to laminar flow conditions that are common in microfluidics. First of all, in reaction chamber volumes from ten to a few hundred microliters, where the dimensions are too large for diffusion to be effective and mixing by convection is slow because of a lack of turbulence. In this case, it becomes difficult to mix the cultivation broth rapidly. Here, diffusion alone is not sufficient for rapid mixing in *MBRs* (Karnik, 2015). Secondly, it is crucial to provide an efficient oxygen supply to satisfy the oxygen uptake rate (*OUR*). Due to the low oxygen solubility in aqueous cultivation media and the high oxygen consumption from aerobic bioprocesses, the supply of oxygen characterized by the oxygen transfer rate (*OTR*) to microorganisms is the most important transport process and may lead to mass transfer limitations.

To overcome these challenges, a microbubble column-bioreactor (μBC) concept for biotechnological research has been developed and characterized. Bubble columns are reactors in which a discontinuous gas phase in the form of bubbles moves relative to a continuous liquid phase. The stream of bubbles enables sufficient aeration and at the same time ensures the homogenization of the cultivation broth. The buoyancy of the bubbles and the momentum exchange between air and liquid promote unsteady liquid flows, which are essential for mixing of the liquid phase. Thus, the main advantages of using bubble columns instead of other multiphase reactors are higher mass and heat transfer rates and less maintenance due to the absence of moving parts.

Previous research was performed within the research unit FOR 856 *Microsystems for particulate life science products* (*mikroPART*, 2011-2014) at TU Braunschweig with a μBC for aerobic cultivation processes of the model organism *Saccharomyces cerevisiae* (Krull and Peterat, 2016; Peterat et al., 2014). The μBC developed in the *mikroPART*-project consisted of two components: a glass substrate and a patterned polydimethylsiloxane (*PDMS*) chip, which was fabricated by soft lithography technology. To prevent adhesion of cells and air bubbles on the hydrophobic *PDMS* reactor walls, the μBC had to be additionally hydrophilized.

However, the aim of the present PhD thesis is focused on the μBC redesign with a reaction volume of 60 µL made exclusively of borosilicate glass, to overcome the existing challenges of working with *PDMS*, and to minimize poorly mixed reactor regions. Using borosilicate glass as manufacturing material avoided the hydrophilization step, and at the same time, it maintained the optical transparency and biocompatible properties of *PDMS*.

The 60 µL μBC will be characterized in detail for the mixing and mass transfer performances. Here, the influence of the superficial gas velocity on the volumetric liquid-phase mass transfer coefficient will be shown as well as the influence on other mass transport-related parameters, e.g., gas hold-up, Sauter mean bubble diameter, bubble rise velocity, superficial liquid velocity, volumetric power input, and mixing time. Additionally, a simplified computational fluid dynamic (*CFD*) model will be developed as a complement to the experimental research. The *CFD* model served as a supporting numerical tool to estimate the fluid dynamics inside the redesigned *MBR*.

To further develop the *μBC* concept, a second *MBR* will be presented in this thesis. It is a custom-made microbioreactor, slightly bigger (550 μL) than the borosilicate glass *MBR*, and it is made of polystyrene, facilitating the manipulation for the sensor integration. The integration of miniaturised optical and electrochemical sensors will allow the real-time and online monitoring of the main cultivation process parameters. As a demonstration, example batch cultivations of *Saccharomyces cerevisiae* will be performed. Validation through these batch cultivations proved the long-term functionality of the sensors and reactor and established that process variable evolution could be observed over time.

In detail the aim of the study is:

- The design and development of a *MBR* for biotechnological research,
- the characterization of the *MBR* and investigation of the mixing performance and mass transfer characteristics,
- the development of a simplified computational fluid dynamic model as a complement to the experimental research,
- the integration of miniaturised sensors for optical density, dissolved oxygen, *pH* and glucose in the *MBR* for the real-time online monitoring of bioprocess variables,
- the validation of the *MBR* and proof of the functionality of the sensors with a cultivation of a model microorganism,
- the application of the *MBR* to determine reaction kinetics of other microorganisms and to biocatalysis research.

The thesis consists of eight chapters structured as follows:

Introduction and main goals of the thesis (chapter 1), introduction to microfluidics and to *MBR*s (chapter 2). It presents the state of the art of *MBR*s, highlighting those focused in bacteria cultivations, the challenge of sensor integration and the characterization of *MBR*s. Chapter 2 gives an overview to bacterial cultivations presenting the main operation modes of bioreactors and the used strains, and finally it presents the application of heterogeneous biocatalysis into *MBR*s. Chapter 3 describes the materials and methods used in the whole work. This includes the design of two *MBR*s, the strains and enzymes utilized, and the followed methods to characterize the developed *MBR*s, to cultivate and perform biotransformations into them, and the description of the integrated miniaturised sensors. Chapter 4 will show the characterization of a glass-based *MBR*. Here, a detailed engineering characterization including the mixing performance and oxygen mass transfer is given. Additional *CFD* simulations are performed to understand better the behaviour of the fluids in the *MBR*. The system is validated by a batch cultivation of the model organism *S. cerevisiae*. Chapter 5 is focused on the sensor integration for the online monitoring of process variables. While the borosilicate *μBC* includes just the sensors for optical density (*OD*) and dissolved oxygen (*DO*), the integration of sensor spots for *pH* and *DO* and of a glucose biosensor is performed in a second developed *MBR*. The *MBR* and the sensors are validated with a *S. cerevisiae* cultivation. Two main *MBR* application examples will be shown in chapter 6: i) the cultivation in batch and continuous mode of the microorganism *Staphylococcus carnosus* and ii) the biotransformation in batch an continuous mode in a microfluidised bed bioreactor. Chapter 7 gives the main conclusions of the thesis and the outlook for future works. The work concludes with the bibliography (chapter 8).

4

2 Theoretical background

2.1 State of the art of microfluidics and microbioreactors

Research in the field of microfluidics has its origin in the 1970s with the increasing development of microtechnical production processes, but it was not until the 1990s that it became a relevant research topic, especially with the contributions of Manz and Harrison (Harrison et al., 1992; Manz et al., 1990). Microfluidics deals with the behaviour, precise control and manipulation of fluids that are geometrically constrained to a small, typically sub-millimetre scale. Its main application is the design of devices capable of performing fluid manipulations with very low volumes. It has properties that make it attractive as practical and feasible tool in biotechnology areas. In terms of microfluidics, the main forces that operate on fluids and influence its behaviour are diffusion, convection, inertial, viscous, and interfacial forces. But concerning the fluid dynamic conditions in microfluidics, diffusion is the dominant mass transfer phenomenon due to miniaturised channels. Furthermore, in microfluidic channels, the typically high surface area to volume ratio introduces surface tension and surface wettability. The knowledge of physics and fluid dynamics concepts, as well as several possibilities of material and geometry configurations are important aspects for developing new microfluidic designs with proper applications (Oliveira et al., 2016).

Bioprocesses involve the use of microorganisms or their constituent parts such as enzymes, as a catalyst for producing valuable substances such as recombinant proteins, drugs or bio-based chemicals. Conventional methods for bioprocess development use extensive experiments on laboratory scale to select the best cell metabolism conditions to improve the productivity. These screening processes are not easy to parallelise, require time for experimentation and chemical analysis, and a large number of samples, which leads to the generation of waste. Consequently there is a great need for high-throughput devices that allow rapid and reliable bioprocess development, for instance microbioreactors (MBRs). MBRs are miniaturised bioreactors with a working volume below 1000 µL (Kirk and Szita, 2013). The origins of MBRs lay in the miniaturised total chemical analysis system (TAS), also known as lab-on-a-chip (LoC). During the early 2000s, there were several contributions in the field of miniaturised reactors (Betts et al., 2006; Doig et al., 2005a; Doig et al., 2005b; Kostov et al., 2001; Lamping et al., 2003; Puskeiler et al., 2005). *MBRs* were introduced later by the Jensen group (Szita et al., 2005; Zanzotto et al., 2004; Zhang et al., 2006) and have been developed as a screening tool for bacterial and mammalian/ human cell cultivation systems, especially for biotechnological, pharmaceutical and medical process development and optimization (Krull and Peterat, 2016; Schäpper et al., 2010). The main research foci include the feasibility of online analytic integration and strain-dependent biological reaction kinetic parameter measurement.

The main advantages of *MBRs* are the minimization of space, reagents, their easy parallelization that together with the integration of analytical tools makes them very interesting devices to develop bioprocesses with *HTS* potential (Krull et al., 2016; Marques and Szita, 2017). *MBRs* are suitable tools for screening applications, e.g., pharmacokinetics, drug delivery or metabolic flux studies, where expensive and only limited amounts of agents are used, or when screening and analysis of dangerous substances (e.g., gas fermentations) is intended. Therefore, the employment of *MBRs* represents a significant step to accelerate bioprocess research, optimising the cultivation conditions and biocatalytic processes, and consequently increasing productivity at large scale.

The scalability of the *MBR* to the macroscale and *vice versa* is reproduced and validated in several cases, showing their potential to provide much of the data and functionality that a large bioreactor system makes available while offering the advantages of high-throughput processing (*HTP*) in terms of costs, space, and time (Krull et al., 2016). Other advantages that *MBR* offer are automation and standardization of the processes (less manual work), high velocity and resolution of analysis with high information content, high degree of parallelisation, rapid process optimization, portability, ease of manipulation, and minute samples.

MBRs have many advantages, but it is important to keep in mind that they could also present some drawbacks related to laminar flow conditions that are common in microfluidics. In reaction chamber volumes from ten to a few hundred microliters, where the dimensions are too large for diffusion to be effective and mixing by convection is slow because of a lack of turbulence, it becomes difficult to homogenize the cultivation broth rapidly. Rapid and efficient mixing is a fundamental requirement for proper distribution and suspension of the substrate and microorganisms during growth cultivation to ensure a uniform process performance and the quality and reproducibility of the measurements. In macroscale systems, the effect of inertia is often significant, resulting in large Reynolds numbers and turbulence, which can be beneficial for mixing. However, due to the small dimensions, microfluidic flows are characterized by low Reynolds numbers, typically in the range of 0.01 – 100, and the effects of inertia are often negligible (Karnik, 2015). Turbulence is therefore typically not encountered, and mass transport is therefore diffusion-limited.

The high impact of *MBRs* on biotechnology and medical applications is reflected in the number of publications involving *MBR* studies of bacteria, yeast, human and mammalian cells, and biocatalysis over the last 30 years are shown in Fig. 2–1. The search was conducted using the web Scopus from Elsevier, and the number of publications corresponds to the amount of accumulated publications over time and the symbol (*) took into account permutations of the keyword. The results of the literature search revealed the increasingly important role of *MBRs* in biotechnological research.

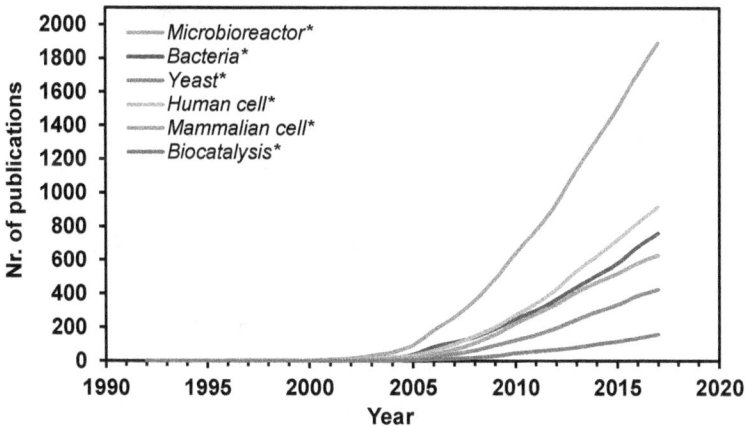

Fig. 2–1 Number of microbioreactor publications over the last 30 years concerning studies with "Microbioreactor*" as a search term along with: "Bacteria*", "Yeast*", "Human cell*", "Mammalian cell*," and "Biocatalysis*".

2.1.1 State of the art of microbioreactors for bacterial and yeast cell cultivations

The main focus of this work is the development of MBRs for bacterial and yeast cell cultivations. Current research in this field is very active, proof of which is the constant publication of reviews concerning its state of the art. Schäpper et al. (Schäpper et al., 2009) reviewed MBRs for cultivation in suspension, focusing on their fabrication materials, mass and heat transfer issues, sensing and control details, and industrial applicability. Marques et al. (Marques et al., 2010) reviewed the criteria for scaling-up cultivation/ bioconversion processes from MBR to lab-scale. Gernaey et al. (Gernaey et al., 2012) provided an updated view on the status of MBRs, identifying critical needs and issues for furthering the successful development of MBR monitoring and control. Hegab et al. (Hegab et al., 2013) gave an overview of MBR fabrication techniques as well as their operation and control. Kirk and Szita (Kirk and Szita, 2013) examined oxygen transfer in miniaturised (milliliter-ranged) bioreactors and microliter-ranged MBRs and showed comparable characteristics to bench-, pilot-, and production-scale systems. Lattermann and Büchs (Lattermann and Büchs, 2015) reviewed miniaturised bioreactors and MBR development, highlighting the mass transfer and power input characterization, optical monitoring, and automation of fed-batch screening systems. Krull et al. (Krull et al., 2016) reviewed the MBR platforms used for bacterial and mammalian/ human cell cultivation biotechnology and process development, focusing on the fabrication material, mixing and aeration methods, and implemented sensors of the most relevant MBRs in the literature. Oliveira et al. (Oliveira et al., 2016) introduced the possibilities of microfluidics to be applied in the field of industrial biotechnology, presenting the principal definitions and fundamental aspects of microfluidic parameters to better understand advanced approaches. Ladner et al. (Ladner et al., 2017) reviewed the miniaturised bioreactors and MBR designed to obtain deeper insight on the level of microbial physiology, single-cell micro-cultivation devices to study the impact of microbial phenotypic heterogeneity on bioprocesses. And finally, Marques and Szita (Marques and Szita, 2017) recently published a review in which the main focus was the development of microfluidic devices for the production of small molecules, therapeutic proteins, and cells.

Besides all the reviews presented here, the main *MBRs* for *HTP* applications to screen bacterial and yeast cell cultivations and to optimize their growth and production in planktonic suspension are summarised in Tab. 2–1. Apart from these prototypes developed in research groups, commercial miniaturised systems in the mL-range with integrated sensors have been also developed like the automated *ambr*-system for cell line cultivation (Sartorius, Göttingen, Germany, working volume of 10 - 15 mL, equipped with *pH* and *DO* sensors) and the microtiter plates (*MTPs*) BioLector (m2p-labs GmbH, Baesweiler, Germany, 0.8 - 2.4 mL equipped with sensors for *OD*, *pH*, *DO*, and fluorescence).

Tab. 2–1 Overview of the developed microbioreactor systems: Bacterial and yeast cell cultivations. Adapted from Krull et al., 2016.

Reference	Type	Mode of operation	Volume	Organism	Application	Material	Mixing	Aeration	Implemented sensors
Yamamoto et al., 2002	Micro-reactor array	Batch	8 chambers x 125 nL	Escherichia coli	HTP cell-free protein synthesis	PDMS and glass		Passive aeration via diffusion	T control chip, intensity of GFP and BFP fluorescence
Maharbiz et al., 2003	MBR array	Batch	250 µL	Escherichia coli	Electrolytic gas generation as a key component for aeration in a miniaturised bioreactor array	Plastic microplate strips, commercial printed circuit board, gold wire, silicone membrane	Mixed via external shaker	Passive aeration via diffusion	OD (LEDs) and silicon photodiodes (PDB- C154SM; Photonic Detectors, Simi Valley, USA), pH (ISFET sensor chip, Sentron Europe), T control via buried thick film thermistors and heaters under each well
Peng and Li, 2004	3D flow controlled microchip	Continuous fresh medium	15 µm deep channels with 15 µm radius	Saccharomyces cerevisiae	Cell scanning, and single-cell fluorescent measurement on a single yeast cell, (applicable to mammalian cells)	Glass	3D flow (driven by electric potentials)	Passive aeration through perfusion	Flow field (Optical microscopy, inverted microscope (Nikon TE 300) with a dual-image module (Nikon))
Zanzotto et al., 2004	Membrane-aerated MBR	Batch	5 and 50 µL	Escherichia coli	Reproduce growth kinetics, observed in bench-scale volumes, in glass MBR	PDMS and glass	Diffusion	Passive aeration via diffusion	DO (PSt3, PreSens), pH (HP2A, PreSens), OD (orange LED, Epitex L600- 10V, 600 nm), LEDs and photo- detectors (PDA-55, Thorlabs)
Balagaddé et al., 2005	Micro-chemostat	Continuous	6 x 16 nL	Escherichia coli	Monitor the prgrammed behaviour of bacterial populations during long cultivation time	PDMS	Diffusion	Passive aeration via diffusion/ influx of fresh medium	Optical microscopy (Nikon TE 2000, Heinze)
Groisman et al., 2005	Micro-chemostat in an array	Continuous	340 µL chambers 100 µm x 70-200 µm x 6 µm	Candida albicans, E. coli, S. cerevisiae, Staphylococcus epidermidis, Pseudomonas aeruginosa	Grow cell colonies to high density starting from one cell, and monitor them for extended time at a single-cell resolution	PDMS and glass	Through the influx of fresh medium	Passive aeration via diffusion	Colony growth (series of fluorescence images), T (infrared camera)
Szita et al., 2005	Multi-plexed MBR	Batch	150 µL	Escherichia coli	Demonstrate the reproducible performance of the multiplexed system	PMMA and PDMS	Magnetic spin bar	Passive aeration via diffusion	Cell density and morphology with single-cell resolution (Optical microscopy, Nikon TE 2000, Heinze)
Zhang et al., 2005	MBR with integrated optical sensors	Batch	150 µL	Escherichia coli; Saccharomyces cerevisiae	Reproducible growth kinetics observed in conventional experiments	PMMA and PDMS	Ring-shaped magnetic stir bar	Passive aeration via diffusion	DO (PSt3, PreSens), pH (HP2A, PreSens), OD (LED, Epitex L600-10V, 600 nm), LEDs and photodetectors (PDA-55, Thorlabs)

Tab. 2–1 Continuation

Reference	Type	Mode of operation	Volume	Organism	Material	Mixing	Aeration	Implemented sensors	Application
Boccazzi et al., 2006	MBRs equipped with internal stirring	Batch	150 µL	Saccharomyces cerevisiae	PMMA and PDMS	Ring-shaped magnetic stir bar	Passive aeration via diffusion	DO (PSt3, PreSens, Germany), pH (HP2A, PreSens), OD (orange LED, Epitex L600- 10V, 600 nm), LEDs and photodetectors (PDA-55, Thorlabs)	Demonstrate the reproducible performance of differential gene expression analysis
Lee et al., 2006	MBR array with integrated mixers	Batch	100 µL	Escherichia coli	PDMS	Peristaltic oxygenating mixer	Peristaltic oxygenating mixer	pH control (injecting base or acid into the growth wells), DO control (varying the oxygen concentration of the peristaltic oxygenating mixer actuation gas), OD	Provide a platform for the study of the interaction of microbial populations with different environmental conditions
Zhang et al., 2006	Micro-chemostat	Continuous	150 µL	Escherichia coli	PMMA and PDMS	Ring-shaped magnetic stir bar	Membrane-aerated	DO (PSt3, PreSens), pH (HP2A, PreSens), Thermocouple (TP- 2444, TE Technology), T control (TC-24-10, TE Technology), OD (LED, Epitex L600- 10V, 600 nm), LEDs and photodetectors (PDA-55, Thorlabs)	Microchemostat as effective tool for investigations of cell physiology and metabolic rates
Zhang et al., 2007	MBRs integrated with automated sensors and actuators	Batch, fed-batch, continuous	150 µL	Escherichia coli	PMMA and PDMS	Ring-shaped magnetic stir bar	Passive aeration via diffusion	DO (PSt3, PreSens), pH (HP2A, PreSens), LEDs and photodetectors (PDA-55, Thorlabs.)	Demonstrate the feasibility of culturing microbial cells, in MBRs
Edlich et al., 2010	Diffusion-based MBR	Continuous	8 µL	Saccharomyces cerevisiae	PDMS and glass	Diffusion	Passive aeration via diffusion	DO (PreSens), OD (orange LED and a photosensor)	Work in parallel and continuous modes to study the growth kinetics of S. cerevisiae
Schäpper et al., 2010	Aerated single- use MBR	Batch, continuous	100 µL	Escherichia coli; Saccharomyces cerevisiae	PDMS	Small magnetic stirrer bar	Diffusion	OD (yellow LED, L600-10 V, Epitex and NT45-317 Edmund Optics), DO (sensor spot, SP-PSt3-NAU-D4-YOP, PreSens), pH (blue LED, 465 nm, NSPB500S, Nichia Corporation), T (Pt100 sensor, JUMO)	Research and development in a screening and in a production environment with a cheap, easy to use and disposable MBR

11

Tab. 2–1 Continuation

Reference	Type	Mode of operation	Volume	Organism	Application	Material	Mixing	Aeration	Implemented sensors
Demming et al., 2011a	Parallel MBR with grid structure engraved on each chamber	Batch, continuous	5 chambers x 9 µL	Aspergillus ochraceus	Monitor the different germination behaviours of submerged cultivated spores	PDMS and glass		Passive aeration via diffusion	T controlled incubation chamber, T controller (KT4, Panasonic), switch cabinet heating system (RO/SE, Bad Bimbach) and Peltier element (Quick-Cool QC 127-1.4-8.SMD, Quick-Ohm Küpper), MBR located under an optical microscope (Axioskop, Zeiss)
Demming et al., 2011b	MBRs with segmented waveguides	Continuous	11 µL	Saccharomyces cerevisiae	Local absorbance measurement and continuous cell monitoring. Study of either the spatial or the temporal evolution of a given analyte.	PDMS and glass		Passive aeration via diffusion	T (heating plate consisting of a foil heater (Minco Europe GmbH) and a Pt-100 thermo couple (Heraeus Sensor Technology), optical characterization through the microspectrometer HR4000 (Ocean Optics, Dunedin)
Lee et al., 2011	Micro-fluidic chemostat	Batch, continuous (chemostat and turbidostat)	1000 µL	Escherichia coli	Characterize production of acids/ products and yields. Characterization of cell dynamics or induce chemically dependent responses	PDMS and plastic	Peristaltic mixer	Passive aeration via diffusion	pH sensor spots (Presens) Oxygen sensor, spots in the base of the growth chamber sections, T control (closed loop PID), OD (LEDs and fibers)
Demming et al., 2012	Vertical micro-bubble column (µBC)	Batch	70 µL	Saccharomyces cerevisiae	Validate the scalability from lab- scale to microscale, and demonstrate its application as screening instrument	PDMS and glass	Mixing thanks to the generation of microbubbles	Rising microbubbles of air or oxygen	OD with integrated optical fiber sensors (Ocean Optics), bubble formation, T (integrated heating structures made of indium tin oxide)
Moffitt et al., 2012	Single-cell chemostat based in agarose	Continuous	1 µL	Bacillus subtilis, Escherichia coli, Enterococcus faecalis	Study the natural heterogeneity in growth and gene expression	PDMS	Diffusion	Passive aeration via diffusion	Bacterial growth (through optical microscopy with a home-built phase- contrast microscope and SEM, Supra55VP, Zeiss)
Grünberger et al., 2012; Grünberger et al., 2013	Picolitre bioreactor for single-cell analysis	Continuous	1 pL	Corynebacterium glutamicum, Escherichia coli	Cultivation of bacteria on single-cell level to study population heterogeneity	PDMS	Diffusion	Passive aeration via diffusion	Nr. cells and cells size (DIC microscopy images and fluorescence images, Nikon NIS Elements AR software package), production studies

Tab. 2–1 Continuation

Reference	Type	Mode of operation	Volume	Organism	Application	Material	Mixing	Aeration	Implemented sensors
Long et al., 2013	Micro-fluidic chemostat	Continuous	600 channels 13-20 µL	Escherichia coli	Growth dynamics (growth rate, cell sizes and GFP expression) at the single-cell level	PDMS	Diffusion	Passive aeration via diffusion	Growth rate and GFP expression (Optical microscopy with inverted microscope, Leica DMI-4000B)
Park et al., 2013	Micro-chemostat array	Continuous perfusion	Array 8 x 250 nL	Saccharomyces cerevisiae	Study and monitor metabolic profiling/ screening, drug response of microorganisms or conduct parallel testing of competing microbes	PDMS	Diffusion	Passive aeration via diffusion	Cell density (microscopic images, Nikon America)
Kunze et al., 2014	MTP for HTP temperature profiling	Batch	96 wells x 200 µL	Escherichia coli, Kluyveromyces lactis	Tool for rapid characterization of temperature dependent reaction processes	96-well lumox multiwell plates (Greiner Bio-One)	Mixed via external shaker	Aeration by pressurized air	Online monitoring with BioLector (m2p-labs), T (via fluorescent dyes a mixture of Rhodamine B (RhB) and Rhodamine 110 (Rh110))
Peterat et al., 2014; Krull and Peterat, 2016	Vertical micro-bubble column (µBC)	Batch	70 µL	Saccharomyces cerevisiae	Characterization of mass transfer, chemostat cultivation, growth kinetics	PDMS and glass	Mixing thanks to the generation of micro-bubbles	Rising micro-bubbles of air and passive aeration via diffusion	OD (HL-2000-HP-FHSA, Ocean Optics), DO (NTH-PSt1-L2.5-TF-NS40/0.4-OIW, Presens), T (custom-made incubation chamber)
Wilming et al., 2014	Disposable MTP system	Fed-batch	44 wells x 40-200 µL	Escherichia coli, Hansenula polymorpha	Primary screening of strains, media, and process conditions at an early stage of process development	PMMA	Mixed via external shaker		Online monitoring with BioLector (m2p-labs)
Probst et al., 2015	Parallel picolitre cultivation chambers	Continuous	1000 µL chambers 40 - 50 µm x 60 µm x 1 µm	Corynebacterium glutamicum	Fast transfer from pre-culture to the main microfluidic chip culture using an entrapped nanolitre sized air bubble	PDMS	Diffusion	Passive aeration via diffusion	Particle trajectories and cell growth (inverted time-lapse microscope, Nikon TI-Eclipse, Nikon)

13

Tab. 2–1 Continuation

Reference	Type	Mode of operation	Volume	Organism	Application	Material	Mixing	Aeration	Implemented sensors
Kirk et al., 2016	Oscillating jet driven, membrane-aerated MBR	Batch	<100 µL	Saccharomyces cerevisiae	Speed up the development and optimization of cultivation processes	PDMS	Stirrer bar	Passive aeration via diffusion	DO (PSt3 sensor spots, PreSens), OD (orange LED, LV600-06V, 600 nm, Epitex)
Bolic et al., 2016	Flexible well-mixed MBR	Batch	500-2200 µL	Lactobacillus paracasei, Saccharomyces cerevisiae	Yeast cells and bacteria cultivations	PMMA	Magnetic stirrer	Sparging and surface aeration	T sensor, heater and three custom made optical fibers (for pH, DO and OD)
Brás et al., 2016	Cell trapping MBR	Continuous	200 µm x 100 µm x 9 µm	Saccharomyces cerevisiae	Continuous production of a biomolecule	PDMS	Through the influx of fresh medium	Passive aeration via diffusion and perfusion	Number of cells variation obtained by micrographs

14

Previous *MBR* research was performed at the Institute of Biochemical Engineering (*ibvt*), TU Braunschweig. The *MBRs* developed at the *ibvt*, together with those presented in this thesis are summarized with their main characteristics in Tab. 2–2.

Tab. 2–2 State of the art of microbioreactors at Institute of Biochemical Engineering (*ibvt*), TU Braunschweig

Microbioreactors at *ibvt*	*PDMS/* glass h*MBR*	*PDMS/* glass µ*BC*	Glass-based µ*BC* (gµ*BC*)	Cuvette-based µ*BC* (cµ*BC*)
Reported in	Edlich et al., 2010	Krull and Peterat, 2016; Peterat et al., 2014; Demming et al. 2012	Lladó Maldonado et al., 2018	Lladó Maldonado et al., 2019a; Lladó Maldonado et al., 2019b
Material	PDMS/ glass	PDMS/ glass	Borosilicate glass	Polystyrene cuvette/ polycarbonate slide
Reactor volume	8 µL	60 µL	60 µL	550 µL
Manufacturer	*IMT**	*IMT**	Micronit	*ibvt*
Technology	Soft lithography	Soft lithography	Powder blasting/ wet etching	Custom-made
Integrated sensors				
Optical density (*OD*)	Photosensor (custom-made)	Mini-spectrometer (Ocean Optics)	Mini-spectrometer (Ocean Optics)	Mini-spectrometer (Ocean Optics)
Dissolved oxygen (*DO*)	Sensor spot (PreSens)	Microneedle sensor (Presens)	Microneedle sensor (Pyroscience)	Microneedle sensor, (Pyroscience)/ Sensor spot (TU Graz)
Glucose	-	-	-	Biosensor (U. Oulu)
pH	-	-	-	Sensor spot (TU Graz)
Mass transfer capacity (k_La)	115 1/h	504 1/h	320 1/h	775 1/h
Mixing time		1.4 - 14 s	5.5 - 13 s	< 1 - 3 s
Application	Cultivation	Cultivation	Cultivation	Cultivation/ Biocatalysis
Operation mode	Continuous	Batch/ Continuous	Batch	Batch/ Continuous
Strain	S. cerevisiae	S. cerevisiae	S. cerevisiae	S. cerevisiae/ S. carnosus
Enzyme	-	-	-	Glucose oxidase

*IMT: Institute of Microtechnology, TU Braunschweig

Edlich et al. (Edlich et al., 2010) developed a horizontal diffusion-based microbioreactor system (hMBR) operated with a reaction volume of 8 µL. The hMBR was made of glass and PDMS manufactured by soft lithography technology, and it had integrated OD and DO sensors. The oxygen diffused into the cultivation broth through the PDMS membrane. However, the disadvantage of the hMBR was that bubbles that occur could remain in the system, thus, displacing the liquid, influencing or completely blocking the liquid flow and disturbing optical measurements.

An alternative and improved operation was a vertical configuration with active gassing developed by Demming and Peterat (Demming et al. 2012; Krull and Peterat, 2016; Peterat et al., 2014). It consisted of a PDMS-glass-based microbubble column-bioreactor manufactured by soft lithography technology with a working volume of 60 µL. Because the naturally hydrophobic PDMS caused problems, such as adherence and entrapment of cells and air bubbles on the unmodified reactor walls, the fabrication was finalized with a hydrophilization (Schmolke et al. 2010). The system was used for aerobic submerged chemostat cultivation and compared with data generated on macroscale stirred tank reactors, obtaining a good agreement.

To overcome the existing challenges of working with PDMS, a new designed µBC was developed and presented in this thesis in chapter 4. It consisted of a borosilicate glass-based microbubble column-bioreactor (gµBC) with a working volume of 60 µL (Lladó Maldonado et al., 2018). In this work Lladó Maldonado et al. developed a gµBC manufactured by powder blasting and wet etching technology. The active aeration of the gµBC allowed proper aeration of the cultivation broth, while ensuring its homogenization and preventing cell sedimentation. The gµBC was equipped with online sensors for optical density (OD) and dissolved oxygen (DO) using a microneedle oxygen sensor.

To facilitate the integration of miniaturised sensors, another prototype was developed. It consisted of a custom-made cuvette-based microbubble column-bioreactor (cµBC) with a reaction volume of 550 µL (Lladó Maldonado et al., 2019a; Lladó Maldonado et al., 2019b). The increase in the volume facilitated the integration of the sensors. The results are presented in chapters 5 and 6.

2.1.2 Microreactors for biocatalysis

MBRs are not just useful for screening of microorganisms; they are also an increasingly used tool in biocatalysis research. Besides the screening of suitable enzymes for a certain process in MTPs or MBRs (Gruber et al., 2017b), it is also considered as a replacement for bigger scaled batch processes since especially continuous (plug-flow) reactors with wall-immobilized enzymes exhibit advantageous properties for enzymatic conversion like improved process control (Bolivar and Nidetzky, 2013; Matosevic et al., 2011; Thomsen and Nidetzky, 2009). In addition, continuous processes often outplay batch processes due to increased yields and lower energy costs (Marques et al., 2012). The increased productivities at microscale are therefore mainly due to less mass transfer limitations and the high surface-to-volume ratio in microfluidic devices which allows process intensification compared to higher scaled approaches (Bolivar et al., 2016b).

Apart from plug-flow microreactors, falling film microreactors (FFMR) (Bolivar et al., 2016a; Illner et al., 2014) and miniaturised bubble columns (Kojima and Suzuki, 2006; Nahalka et al., 2008) are under investigation for oxygen dependent enzymatic conversions, where the enzymes are immobilized inside the walls or to porous particles, respectively. Immobilization of enzymes is a well-known unit operation of biocatalytic reactions that allows the retention and re-use of the biocatalyst. Both reaction systems exhibit the potential for continuous bioconversion due to their steady oxygen supply with high k_La values compared to for instance systems as shaken flasks. The conducted experiments during this study further evaluated the chances of microbubble columns as an advanced screening tool for oxygen dependent bioconversion at microscale.

2.2 Sensor integration in microbioreactors for monitoring biotechnological process variables

The monitoring of biotechnological key process variables, e.g., carbon source concentration, biomass concentration and growth behaviour, and product concentration, provides essential information for biotechnological process characterization. The online monitoring of these parameters offers real-time analysis and process control and requires no sampling, which reduces the risk of process contamination and undisrupted flow. Conventional lab-scale bioreactors (with reaction volumes of 0.25 to 5 L) are usually fully equipped with sensors, but when down-scaling to *MBRs* the integration of sensors becomes less practical and challenging.

Since it is not possible to insert standard probes because of their dimensions, and since continual sampling for at- or offline-HPLC measurements would rapidly deplete the cultivation medium of a *MBR*, integration of miniaturised sensors are typically preferred for monitoring the process conditions (Ehgartner et al., 2016; Gruber et al., 2017a; Lasave et al., 2015; Panjan et al. 2018; Pfeiffer and Nagl, 2015). They offer online and real time monitoring of several bioprocess variables during cultivations or enzymatic bioconversions, reduce the need for sampling, avoid sample loss and contamination, and thus significantly reduce laboratory effort.

Optical chemical sensors are an attractive format for integration into microfluidic devices due to the following features: they are highly sensitive, inexpensive, and easy to miniaturise; and they are capable of non-invasive and non-destructive monitoring during operation (Ehgartner et al., 2016; Gruber et al., 2017b). Several reviews are published on the integration and application of optical sensor in microfluidic devices (Gruber et al., 2017a; Pfeiffer and Nagl, 2015; Sun et al., 2015). Alternatively, electrochemical biosensors provide an attractive option for the analysis of biological sample content due to their direct conversion of a biological response to a quantifiable and processable electronic signal. The advantages of electrochemical biosensors include their robustness, easy miniaturisation, and excellent detection limits even with small analyte volumes, as well as their ability to be used in turbid cultivation broths with optically absorbing and fluorescing compounds (Grieshaber et al., 2008).

2.3 Fluid dynamic characterization of microbioreactors

The aim of characterising novel *MBRs* is to demonstrate the technical feasibility of the developed microdevice for screening of cultivations, with special emphasis to the factors affecting the mixing performance and oxygen mass transfer. The *MBR* concept developed here is a *µBC*, therefore it is relevant to proof that the stream of bubbles provides a sufficient aeration to satisfy the high demand of oxygen of aerobic bioprocesses and at the same time enables the homogenization of the cultivation broth. For this reason, it is investigated the influence of the superficial gas velocity (u_G) on the volumetric mass transfer coefficient (k_La) and on other important mass transport relating parameters like gas hold-up (ε_G), Sauter mean bubble diameter (d_{vs}), bubble rise velocity (u_b), superficial liquid velocity (u_L), volumetric power input (P/V). Additionally, computational fluid dynamics (*CFD*) modelling is employed in this research as a supporting numerical tool to estimate the fluid dynamics inside the μBC.

Some examples of engineering characterized reactors in the millilitre scale for microbial cultivations could be found in the literature. Doig et al. (Doig et al., 2005b) described the characterization of a 2 mL-bubble column-bioreactor, showing the influence of the superficial gas velocity (u_G) on the volumetric liquid-phase mass transfer coefficient (k_La) and a correlation between power consumption and k_La. Betts et al. (Betts et al., 2006) performed power input, mixing time, and k_La characterization of a 10 mL-miniaturised stirred bioreactor that allowed a successful scale-down from a 7 L-stirred tank reactor based on equivalent power input, reaching comparable growth and productivity kinetics. Weuster-Botz et al. (Weuster-Botz et al., 2001) studied the oxygen transfer, power input, and mixing times in bubble columns with a working volume of 200 mL. Gernaey's group characterized the oxygen transfer, mixing time and residence time distribution of a millilitre-scale bioreactor system (0.5 – 2 mL) (Bolic et al., 2016) with a high level of flexibility in terms of volume, aeration and mixing.

The bubble population, the gas hold-up (ε_G) and the bubble rise velocity (u_b), together with the heat and mass transfer coefficients, have a significant impact on the hydrodynamics (Kantarci et al., 2005). Many studies have focused on gas-liquid hydrodynamics and mass transfer in microreactors, but most of them on flows in closed microchannels (Dietrich et al., 2013; Khoshmanesh et al., 2015; Niu et al., 2009; Shao et al., 2009) or open-channel devices, such as falling-film microreactors (Sobieszuk et al., 2012). Although it was possible to find hydrodynamic studies on horizontal micro-scale bubble columns (Haverkamp et al., 2006), there was a lack of studies on the hydrodynamics in vertical micro-scale bubble columns. These latter microdevices have contacting analogies to conventional bubble columns, making them interesting for the scale-up/down processes. The average bubble size in a bubble column of large scale has been found to be affected by gas velocity, liquid properties, gas distribution, operating pressure and column diameter (Deckwer, 1992; Kantarci et al., 2005), making all these parameters a focus of study during the characterization of this type of multiphase reactor.

<u>Oxygen transfer</u>

During an aerobic bioprocess, the oxygen is transferred from a rising gas bubble into the liquid phase and then to the site of oxidative phosphorylation inside the suspended cell or enzyme, which can be considered as a solid particle (Garcia-Ochoa and Gomez, 2009). Suspended planktonic microorganisms, which do not form cell agglomerates (e.g., flocs or mycels) show no diffusional resistances or other significant transport limitations for oxygen. Mass transfer at the interface medium/microorganism is high as compared to that at the gas/liquid interface. The resistance at the boundary cultivation medium/microorganism can be neglected in nearly all cases. The simplest and most used theory on gas–liquid mass transfer is the two-film model, and usually the gas–liquid mass transfer rate is modelled according to this theory. Taking into account that oxygen is poorly soluble in water, it is commonly accepted that the greatest resistance for mass transfer is on the liquid side of the interface and that gas- and solid-phase resistances can usually be neglected. Perfect mixing conditions are assumed. This assumption allows use of a single oxygen concentration measurement point to represent the oxygen concentration in the liquid, which applies reasonably well at microscale.

The oxygen content inside the μBC is the result of the continuous oxygen transfer from the air bubbled into the cultivation medium (oxygen transfer rate, OTR) and the oxygen consumption by the cells growing inside the medium (oxygen uptake rate, OUR), as described in eq. (2-1):

$$\frac{dc_L}{dt} = OTR - OUR \qquad (2\text{-}1)$$

Through the two-film theory with a dominance of liquid mass transfer resistance, the oxygen transfer rate (OTR) is given by eq. (2-2):

$$OTR = \frac{dc_L}{dt} = k_L a \, (c_L^* - c_L) \tag{2-2}$$

The liquid film thickness, the diffusion coefficient and the total interfacial gas-liquid area per unit volume are captured by the process parameter $k_L a$ (Garcia-Ochoa and Gomez, 2009; Kirk and Szita, 2013), c_L^* is the oxygen saturation concentration (0.235 mmol/L at 30 °C), and c_L is the oxygen concentration in the liquid phase. The maximum oxygen transfer rate (OTR_{max}) is reached when the oxygen concentration in the liquid phase c_L is 0, calculated according to eq. (2-3):

$$OTR_{max} = k_L a \cdot c_L^* \tag{2-3}$$

The OUR is given by eq. (2-4):

$$OUR = sOUR \cdot c_{CDW} \tag{2-4}$$

where c_{CDW} is the biomass concentration and $sOUR$ the specific OUR.

To supply sufficient oxygen to the cells and to avoid oxygen depletion, the maximum OTR (OTR_{max}) (estimated by the product $k_L a \cdot c_L^*$) must be bigger than the OUR. The OUR was determined from the OTR by using the previously determined $k_L a$ and the DO profile in the liquid phase measured during the course of the cultivation and the values of the derivative of DO versus time curve (Garcia-Ochoa et al., 2010). OUR during the process was obtained from eq. (2-5):

$$OUR = k_L a \cdot (c_L^* - c_L) - \frac{dc_L}{dt} \tag{2-5}$$

Mixing performance

One of the challenges of MBRs is achieving short mixing times. Mixing under laminar flow conditions occurs by diffusion. In the μBC, the rising bubbles induce circular convection in the liquid due to drag forces and thus enhance mixing by reducing diffusion distances and preventing cell sedimentation. The bubble-induced convection ensures adequate mixing performance.

The mixing time (θ) is an essential process parameter because homogenous conditions are required for a good distribution of the nutrients and cells during cultivation processes. A rapid and efficient mixing during substrate addition and cultivation is the basis for the reproducibility of any bioprocess study. Working at microscale leads to different dominant physical phenomena due to the high area-to-volume ratio. Inertial forces that typically result in turbulence and good mixing at macroscale are weak in microfluidics, while surface tension and capillary forces become dominant at microscale.

Fluid dynamics

The Sauter mean bubble diameter (d_{VS}) refers to a sphere diameter with the same volume-to-surface ratio as the bubble size distribution and is calculated by eq. (2-6):

$$d_{VS} = \frac{\sum_{j=1}^{n} N_j \cdot d_j^3}{\sum_{j=1}^{n} N_j \cdot d_j^2} = \frac{6 \cdot V_G}{S_{G,\,tot}} \qquad (2\text{-}6)$$

where N_j is the number of bubbles with a diameter of d_j and $S_{G,tot}$ represents the total surface of bubbles. Gas hold-up (ε_G) is the volume fraction of the dispersed gas phase, and can be calculated by eq. (2-7), where V_G describes the gas volume and V_{tot} the total reactor volume. ε_G depends on u_G, the liquid velocity, the viscosity, the aspect ratio of the column and the gas distributor. However, in all cases, it is an increasing function of u_G, except eventually at the transition from dispersed to coalesced bubble flow regime, where a contraction may be observed.

$$\varepsilon_G = \frac{V_G}{V_{tot}} \qquad (2\text{-}7)$$

Buoyancy and drag forces regulate the bubble's rise velocity in a liquid column. These forces are strongly dependent on fluid properties, gravity as well as d_{VS}.

The Reynolds number in microdevices is small, often on the order of unity or smaller. The Reynolds number of rising bubbles (Re_b) is a function of the liquid density (ρ) (1000 kg/m^3), the bubble rise velocity (u_b), the bubble diameter (in this case the Sauter mean diameter (d_{VS})), and the dynamic viscosity of the liquid (μ) (0.001 kg/(m/s)), according to eq. (2-8).

$$Re_b \equiv \frac{\rho \cdot u_b \cdot d_{VS}}{\mu} \qquad (2\text{-}8)$$

The volumetric power input (P/V) is often used as a scale-up criterion in bioprocesses. Thus, it is important to estimate it and to verify that it is in the same range when compared with different scales. In bubble columns, the P/V is produced by the bubble generation at the nozzle. Because the produced specific surface energy of the gas bubbles is identical to the surface tension (σ) (0.072 N/m), P/V due to gas bubble formation can be calculated with the u_G, the height of the liquid column (H) (0.019 m) and the respective calculated d_{VS} (Weuster-Botz et al., 2001). The estimation of P/V by eq. (2-9) allows for comparison of the performance of the μBC with that of conventional bioreactors:

$$\frac{P}{V} = \frac{6 \cdot u_G \cdot \sigma}{H \cdot d_{VS}} \qquad (2-9)$$

By knowing the superficial liquid velocity (u_L) (obtained from the computational fluid dynamics (CFD) simulations or Particle Image Velocimetry (PIV)), it is possible to calculate the Reynolds number of the liquid (Re_L) for every u_G, with the liquid density (ρ) (1000 kg/m^3), the hydraulic diameter (d_h) of the cross-section of the $g\mu BC$ (1.5 $\cdot 10^{-3}$ m) and the dynamic viscosity of the liquid (μ) (0.001 kg/ (m·s)), according to eq. (2-10):

$$Re_L \equiv \frac{\rho \cdot u_L \cdot d_h}{\mu} \qquad (2-10)$$

Computational fluid dynamics

Flows and transport phenomena can be described by partial differential equations, which cannot be solved analytically. To obtain an approximate solution numerically, it is necessary to use a discretization method which approximates the differential equations by a system of algebraic equations, which can then be solved on a computer (Ferziger and Peric, 2002). Computational fluid dynamics (CFD) is a branch of fluid mechanics that uses numerical methods and algorithms to solve and analyse the behaviour of fluid flows and the effects of fluid motion computationally. CFD has become a widely used method to solve fluid-dynamic problems in addition to experimental methods. Today, the application of CFD is as a suitable tool to determine flow fields, transport phenomena, energy consumption, and substrate uptake, and to identify optimum operation settings for the process investigated (Eslahpazir et al., 2011). CFD is a powerful tool that can be applied to process engineering for improving the design and operation of chemical or biochemical reactors. For example, shape and inlet/outlet locations can be improved through *CFD* simulations to optimize mixing and mass transfer performance.

2.4 Cultivation of microorganisms

Most current *MBRs* operate in batch and fed-batch modes (Bolic et al., 2016; Kostov et al., 2001; Peterat et al., 2014; Schäpper et al., 2010; Zanzotto et al., 2004; Zhang et al., 2005). At these operation modes, the properties of microorganisms, such as size, compositions, and functional characteristics, vary considerably during growth of the cultivation (Marbà-Ardébol et al. 2018). Steady state cell growth, in which cell biomass, substrates and product concentrations remain constant, can only be realized in continuous cultivation experiments. In a chemostat, the cultivation medium is continuously added and removed. Equal influent and effluent flow rates are maintained to retain a constant working volume. As a result, the cells are kept at a steady state with a constant growth rate and metabolic activity. This makes a chemostat an ideal experimental setup to study microbial systems. One of the most important features of the chemostat is that it allows the operator to control the cell growth rate, by adjusting the dilution rate D, which is the ratio of the flow rate divided by the cultivation volume. The major disadvantages of chemostat cultivations are the time required to reach a steady state condition, and the high amount of substrate needed to maintain long cultivation periods, which is prohibitive when expensive substances are used. Some strategies to accelerate the experimental procedures and to reduce the substrate consumption are the use of parallel bioreactors (Klein et al., 2013), and the procedure followed here using the *MBR* (Edlich et al., 2010; Krull and Peterat, 2016; Zhang et al., 2006).

2.4.1 Batch cultivation

Batch cultivation is an operation mode in which cells are grown in a fixed and constant volume of cultivation medium under specific environmental conditions. The phase of microbial growth in a batch culture is generally divided into four phases: lag, exponential, stationary and death phase. The cell growth kinetics in the exponential phase in batch cultivation can be described by eq. (2-11):

$$\mu = \frac{1}{c_{CDW}} \frac{dc_{CDW}}{dt}$$

(2-11)

where μ is the specific growth rate and c_{CDW} the biomass concentration.

2.4.2 Continuous cultivation

Continuous cultivation is an operation mode where the reaction volume is maintained constant through equivalent incoming and outgoing flow rates. This cultivation mode is also known as a chemostat, because it is a stable, self-regulating system with stationary states. Cell growth is ensured by having a constant concentration of a growth-limiting carbon source in the feed flow. The cultivation medium within the reactor is assumed to be ideally mixed; thus, no concentration gradients existed and the concentrations at the outlet corresponded to those within the reactor volume. Substrate-limited growth in a continuous cultivation is a process that allows the investigation of stationary kinetic process parameters.

The biomass balance is:

$$\frac{dc_{CDW}}{dt} = D \cdot (c_{CDW,in} - c_{CDW}) + \mu \cdot c_{CDW} \tag{2-12}$$

where the index "*in*" indicates the corresponding concentrations in the feed flow, and the dilution rate D according to eq. (2-13) represents the quotient of the flow rate F and the reaction volume V:

$$D = \frac{F}{V} \tag{2-13}$$

At steady state $\frac{dc_{CDW}}{dt} = 0$ and $c_{CDW,in} = 0$, so the specific growth rate (μ) corresponds to the dilution rate (D) according to eq. (2-14):

$$\mu = D \tag{2-14}$$

The mass balance for the limiting substrate concentration (c_S) can be expressed as:

$$\frac{dc_S}{dt} = D \cdot (c_{S,in} - c_S) - c_{CDW} \cdot \left(\frac{\mu}{Y_{X/S}} + m_S \right) \tag{2-15}$$

where the yield coefficient $Y_{X/S}$ corresponds to the substrate-related biomass yield coefficient, and m_S is the maintenance coefficient of the endogenous metabolism of the cells.

Using the Monod model, μ can be expressed as in eq. (2-16):

$$\mu = \frac{\mu_{max} \cdot c_S}{c_S + K_S} \tag{2-16}$$

where μ_{max} is the maximum specific growth rate for a theoretically infinitely high substrate concentration and the Monod constant, K_S, represents the substrate concentration corresponding to 1/2 μ_{max}.

The steady state concentrations c_{CDW} and c_S were determined as function of D according to eq. (2-17) and (2-18), respectively:

$$c_{CDW} = \frac{D \cdot (c_{S,in} - c_S)}{\frac{D}{Y_{X/S}} + m_s}$$ (2-17)

$$c_S = \frac{D \cdot K_S}{\mu_{max} - D}$$ (2-18)

When $D \geq D_{washout}$, the biomass would be washed out of the reactor system. The substrate concentration would increase as $D \rightarrow D_{washout}$ and the glucose concentration would reach its input value $c_{S,in}$ at $D_{washout}$, as described in eq. (2-19).

$$D_{washout} = \frac{\mu_{max} \cdot c_{S,in}}{c_{S,in} + K_S}$$ (2-19)

The maximal specific growth rate μ_{max} would theoretically be achieved if the substrate concentration were infinitely large. Therefore, $D_{washout} < \mu_{max}$.

The specific substrate uptake rate q_S can be expressed as in eq. (2-20):

$$q_S = \frac{D \cdot (c_{S,in} - c_S)}{c_{CDW}} = \frac{D}{Y_{X/S}} + m_s$$ (2-20)

By using eq. (2-20), the substrate-related biomass yield coefficient, $Y_{X/S}$, and the maintenance coefficient, m_S, can be determined from the slope and the ordinate intercept, respectively.

The biomass-related productivity (Pr) can be obtained by combining eq. (2-17) and eq. (2-18) and neglecting m_S:

$$Pr = D \cdot c_{CDW} = D \cdot Y_{X/S} \cdot \left(c_{S,in} - \frac{D \cdot K_S}{\mu_{max} - D} \right)$$ (2-21)

The value of D at which the productivity of the cell mass is maximum, defined as $D_{Pr,max}$, is obtained when $dPr / dD = 0$, calculated as follows:

$$D_{Pr,max} = \mu_{max} \cdot \left(1 - \sqrt{\frac{K_S}{K_S + c_{S,in}}} \right)$$ (2-22)

3   Materials and methods

3.1   Materials

3.1.1   Microbioreactor designs and manufacturing

This section describes the two *MBR* prototypes designed and developed during this thesis; a glass-based microbubble column-bioreactor and a cuvette-based microbubble column-bioreactor.

Glass-based microbubble column-bioreactor

The glass-based microbubble column-bioreactor (*gµBC*) consisted of a reaction chamber (3 mm in width, 1 mm in depth, and 18 mm in height) and a funnel at the upper part (5 mm in width, 1 mm in depth, and 14 mm in height) for adequate phase separation (Fig. 3–1 A).

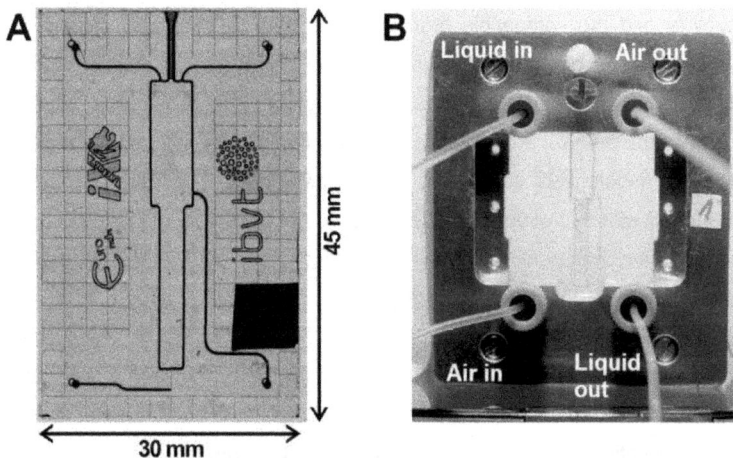

Fig. 3–1 (A) Borosilicate glass-based microbubble column-bioreactor (*gµBC*), and (B) *gµBC* inside the supporting reactor holder (Lladó Maldonado et al. 2018).

The *gµBC* was filled with liquid up to the outlet level (19 mm in height), yielding a final reaction volume of 60 µL. The microdevice consisted of four 30 mm x 45 mm stacked borofloat (*BF*) glass plates (0.5 mm in thickness each), thus being suitable for optical measurements. It was equipped with two inlets and two outlets, one for each phase (gas and liquid). The liquid inlet (used to fill the reactor with cultivation medium) was at the head of the reactor. The liquid outlet was placed at the upper part of the reaction chamber to guarantee a constant reaction volume, and the gas outlet was placed at the top of the funnel. The air was supplied through a nozzle (with a hydraulic diameter of 26 µm) at the bottom of the $g\mu BC$ using the Venturi effect, and before entering the microdevice, it was sterilized by applying filters with a pore size of 0.2 mm. The airflow was accelerated due to a reduction of the cross-section, and therefore pressure energy was transformed into kinetic energy, which is explained by the principle of conservation of energy. The increase in air velocity generated a continuous stream of microbubbles that rose through the vertical device. The motion of the bubbles induced a circular convective flow that enabled the homogenization of the liquid.

The manufacturing of the $g\mu BC$s was carried out in a clean room (Micronit GmbH, Dortmund, Germany) and was divided into two main sections: first, the structuring of the reactor plates, and second, the fabrication of the $g\mu BC$s. Before the fabrication, the reactors and the compatible reactor holder (Fig. 3–1 B) were designed using *CAD* software. The generated design was then transferred to masks, which were written by laser to achieve the required resolution. The two middle plates were structured in the same manner: using standard lithography for etching of *BF* glass, the nozzle channels were transferred to the glass plates. Thus, the channels could be wet etched with hydrofluoric acid (*HF*) with a depth of 10.9 µm into the *BF* glass. Afterwards, the chambers and the connecting channels were structured using standard lithography for powder blasting. The resulting plates then had all required structures for the $g\mu BC$s. The upper *BF* plate of the reactor included holes for access to the channels. The holes were added by standard lithography and powder blasting. The bottom plate remained unstructured. The upper and bottom plates served for closing the functional structures of the reactor. The $g\mu BC$s were built by thermal bonding of the 4 *BF* plates as a sandwich. For accurate performance of the reactor, it was crucial that the alignment was precise. In particular, the nozzle channels needed to be aligned very well to form the desired nozzle with an oval cross-section and the set hydraulic diameter. After the bonding, the reactor chips were separated by dicing. The compatible holder design was passed to a workshop for prototyping the holder in a sandwich structure of aluminum and polymethyl methacrylate (*PMMA*) parts (Fig. 3–1 B).The holder was designed to facilitate the connection of the required tubings and the measurement setup.

Cuvette-based microbubble column-bioreactor

The cuvette-based microbubble column-bioreactor (cµBC) was based on a semi-micro cuvette made of polystyrene (BR759015, Brand, Wertheim, Germany) that was vertically cut in half through the *xy* plane. A poly(methyl methacrylate) (*PMMA*) microscope slide (MS50510415, Labor-und Medizintechnik, Dr. Jutta Rost, Leipzig, Germany) was used to close the vertical open side of the cµBC. A modified needle (Sterican, B. Braun Melsungen AG, Melsungen, Germany) with an outer diameter of 600 µm and an inner diameter manually reduced to less than 100 µm was inserted and sealed at the bottom of the cuvette, which served as a nozzle for the air supply. The cµBC consisted of a reaction chamber (4 mm width, 5 mm depth, and 20 mm height) and a funnel at the upper part (10 mm width, 5 mm depth, and 25 mm height) to ensure an adequate gas/liquid phase separation giving a total volume of 1.5 mL. The cµBC was filled with cultivation medium up to the outlet level, giving a final reaction volume of approximately 550 µL (Fig. 3–2 and Fig. 3–3).

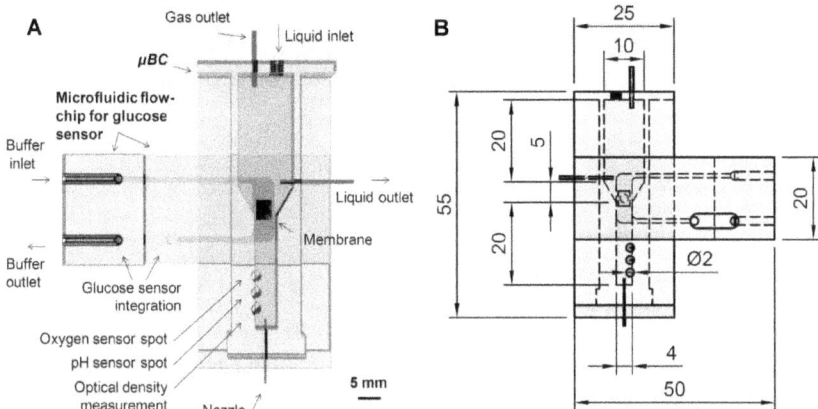

Fig. 3–2 (A) Technical drawing of the cuvette-based microbubble column-bioreactor (cµBC) (front side view) with the microfluidic flow chip for glucose measurement; the inlets and outlets of the liquid and gas phases, and the integrated sensors for *pH*, dissolved oxygen and optical density. (B) Technical drawing of the cµBC (back side view) with dimensions in mm (Lladó Maldonado et al. 2019a).

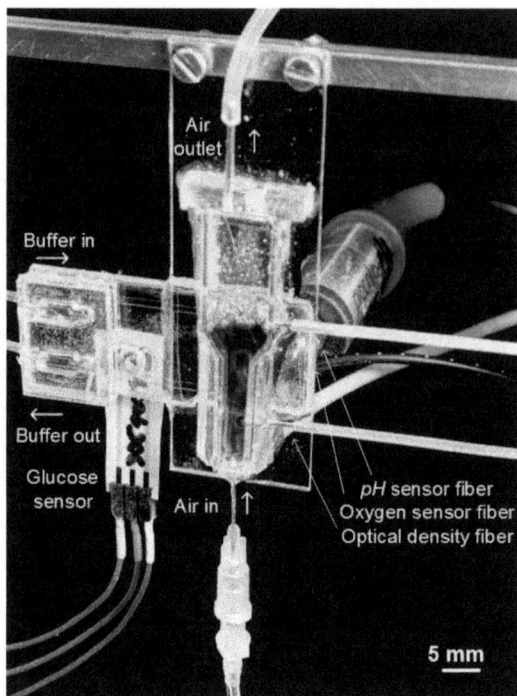

Fig. 3–3 Picture of the cuvette-based microbubble column-bioreactor (*cµBC*) with the microfluidic flow chip and glucose biosensor; the inlets and outlets of the liquid and gas phases; and the integrated sensors for *pH*, dissolved oxygen and optical density with their associated glass fibers.

3.1.2 Strains, enzymes and carriers

Saccharomyces cerevisiae

Saccharomyces cerevisiae is the principal yeast utilized in biotechnology worldwide, due to its key role in many food and other industrial processes. It is also the principal model eukaryotic organism utilized for fundamental research. It can metabolize sugars both aerobically, producing the end products carbon dioxide and water, and anaerobically, producing ethanol and carbon dioxide, as described in eq. (3-1) and (3-2), respectively:

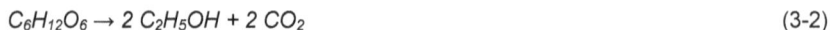

$$C_6H_{12}O_6 + 6\ O_2 \rightarrow 6\ CO_2 + 6\ H_2O \tag{3-1}$$

$$C_6H_{12}O_6 \rightarrow 2\ C_2H_5OH + 2\ CO_2 \tag{3-2}$$

The *S. cerevisiae* strain CCOS 538 was used to validate the *gµBC* and the *cµBC* as aerobic cultivation screening platforms. The *S. cerevisiae* CCOS 538 is a Crabtree-positive strain, a facultatively anaerobic yeast that generates ethanol under aerobic conditions in the presence of excess sugar, under sugar limitation, or when sugar-limited cultivations are suddenly exposed to sugar excess. In the presence of excess of sugar, the amount of glucose that is not metabolized via respiration can be used to produce ethanol via alcoholic fermentation, as described in eq. (3-3). The production of ethanol is also accompanied by the appearance of other metabolites such as acetic acid, pyruvic acid and ethylacetate (van Dijken et al., 1993). For Crabtree-positive *S. cerevisiae* this respiro-fermentative metabolization of glucose has been shown at a glucose concentration of about $\geq$ 150 mg/L, though this might vary from species to species and depend on the specific conditions (Pfeiffer and Morley, 2014).

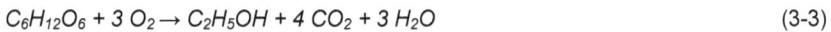

$$C_6H_{12}O_6 + 3\ O_2 \rightarrow C_2H_5OH + 4\ CO_2 + 3\ H_2O \qquad (3\text{-}3)$$

The generated by-product ethanol can subsequently be assimilated in a second growth phase, in diauxic growth, after glucose is completely used and the cells have adapted to the new carbon source (Rieger et al., 1983). During the ethanol-based growth, the ethanol oxidation takes according to eq. (3-4).

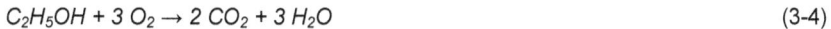

$$C_2H_5OH + 3\ O_2 \rightarrow 2\ CO_2 + 3\ H_2O \qquad (3\text{-}4)$$

The Crabtree-positive *S. cerevisiae* strain CCOS 538 (ATCC 32167) obtained from the Culture Collection of Switzerland AG, was used in this study because it has been well-described and investigated in numerous studies (e.g., see (Beck and von Meyenburg, 1968; Kuhlmann et al., 1984; Rieger et al., 1983; Sonnleitner and Käppeli, 1986) and is thus well suited to be a model organism.

Staphylococcus carnosus

Staphylococcus carnosus as a model organism is a non-pathogenic Gram-positive, facultatively anaerobic, staphylococcal species, which evolved over several decades, from being an important strain in the food engineering industry used as starter culture in sausage production to become a versatile and powerful microbial tool in modern microbiology and biotechnology. The straightforward translocation of recombinant proteins over the single cell membrane in Gram-positive bacteria combined with the very low proteolytic extracellular activity makes *S. carnosus* an attractive host for production of secreted recombinant proteins (Löfblom et al., 2017). Due to its properties, previous studies with the same strain *S. carnosus* were already performed at the microlitre-scale by Davies et al. (Davies et al., 2013) in a *MBR* "cassette".

The strain used was *Staphylococcus carnosus* TM300 GFP containing the plasmid pCX-pp-sfGFP, which carries the information to produce the green fluorescence protein (*GFP*) (Mauthe et al. 2012, Rosenstein et al. 2009, Yu and Götz, 2012).

Carriers

Four different porous particles were used as support for the immobilized enzyme: ReliZyme 403 S (Resindion S.r.l, Italy), Sepabeads EC-EP (Resindion S.r.l, Italy), CPG 300 Å (Biosearch Technologies, USA) and MSU-F (InPore Technologies/DBA Claytec, Inc., USA), and provided by the cooperation working group of Prof. Nidetzky and Dr. Bolivar from the Institute of Biotechnology and Biochemical Engineering (Graz University of Technology, Graz, Austria).

ReliZyme 403 S and Sepabeads EC-EP are polymethacrylate based and have epoxy groups on their surface, whereas CPG 300 Å is a silica material, and MSU-F a mesostructured silica cellular foam with exposed silanol groups, respectively. The geometrical properties of utilised solid supports given in Tab. 3–1 were provided by the designated companies.

Tab. 3–1 Geometrical properties of utilized solid carriers.

Material	Particle size (µm)	Pore size (nm)	Surface area (m²/g)	Pore volume (mm³/g)
Sepabeads EC-EP	100 - 300	10 - 20	-	-
ReliZyme 403 S	100 - 300	40 - 60	-	-
CPG 300 Å	70 - 130	32.18	176.03	1487
MSU-F	3 - 7	17	481	1700

The solid supports were investigated regarding their immobilization yield and final glucose oxidase (GOx) activity. The immobilization yield, which is the percentage ratio of bound and loaded enzyme activity, was determined by measuring the remaining enzymatic activity in the supernatant after the immobilization procedure using the peroxidase-coupled assay. This enabled the calculation of the effectiveness factor η which is the ratio of the observed enzymatic activity and the theoretically immobilized activity calculated from the activity balance (Bolivar and Nidetzky, 2012).

The advantage of MSU-F material to the other investigated enzyme carriers is its high surface area together with the small particle size which enables the loading of a high amount of protein and shorter diffusion paths and is thus less restricted to internal mass transfer limitations as the other materials (Zucca and Sanjust, 2014).

Glucose oxidase

GOx (EC 1.1.3.4) is a dimeric enzyme expressed in *Aspergillus niger* that is classified as an oxidoreductase using molecular oxygen as electron acceptor during its enzymatic reaction (Hatzinikolaou et al., 1996). The catalysed reaction consists of an oxidation of β-D-glucose to D-glucono-1,5-lactone according to eq. (3-5) followed by a non-enzymatic hydrolysis to gluconic acid according to eq. (3-6) and the simultaneous reduction of non-covalently bound flavin adenine dinucleotide (FAD) to $FADH_2$ (Witt et al., 2000).

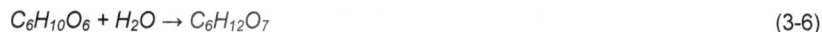

$$C_6H_{12}O_6 + O_2 \rightarrow C_6H_{10}O_6 + H_2O_2 \tag{3-5}$$

$$C_6H_{10}O_6 + H_2O \rightarrow C_6H_{12}O_7 \tag{3-6}$$

GOx (β-D-Glucose:oxygen 1-oxidoreductase, EC 1.1.3.4, type II) from *Aspergillus niger* (Type VII, Sigma Aldrich, Vienna, Austria) was immobilized on the carriers with the method described in chapter 3.2.5.

3.2 Methods

3.2.1 Characterization of a multiphase microbioreactor

Gas flow rates (Q_G) were measured with a custom-made gas flow meter that worked on the basis of liquid displacement, as reported in (Peterat et al., 2014). From the Q_G, the corresponding superficial gas velocities (u_G) were calculated. u_G was calculated as the average velocity of the gas that was sparged into the column, which is expressed as the volumetric flow rate (Q_G) divided by the cross-sectional area of the column (A), as described in eq. (3-7):

$$u_G = \frac{Q_G}{A} \tag{3-7}$$

The fluid dynamic and mass transfer characterization of the µBC was mainly performed by photographic techniques. Special emphasis was placed on studying the air bubbles because they played a fundamental role in the homogenization and mass transfer in the reaction volume. Two-dimensional photography could provide a cross-sectional view of the bubbles. In fact, the µBC was very narrow (1 mm in depth) and the walls were transparent, making it possible to observe the behaviour of the bubbles by selecting the right camera and illumination type. It is important to use a short exposure time (also called shutter speed, it is the length of time that the digital sensor inside the camera is exposed to light) or an ultra-short-duration flash (the duration of light emission per single flash) in relation to the velocity of the bubbles to obtain pictures where the bubbles are in focus and well defined. In this case, the pictures were taken with a flash with a camera (EOS 60D, Canon Deutschland, Krefeld, Germany) with a 90 mm-macro-lens (EF-S 90 mm, Canon Deutschland, Krefeld, Germany). The shutter speed was fixed to 1/10 s, the aperture to 8, and the ISO to 100. Due to the fast movement of the bubbles and to avoid blurred images, a custom-made ultra-short-duration flash ($2 \cdot 10^{-6}$ s) was used. By image analysis of series of captured images taken with the same magnification and same contrast, different parameters were determined for different u_G, the Sauter mean diameter of the bubbles (d_{vs}), the gas hold-up (c_G), and the mixing time (θ).

Mixing time

In the used µBCs the stream of bubbles discharged from the nozzle induced and promoted the mixing of the liquid. Mixing was characterized by introducing a pulse of a fluorescent tracer solution (Fluorescein sodium, Merck, Darmstadt, Germany)

(2 µL and 5 µL of a 50 mg/L solution, in the *gµBC* and *cµBC* respectively) at the inlet of the *µBC* through a precision syringe pump (Nemesys, Cetoni GmbH, Korbussen, Germany). Via analysis of the image sequences captured with an approximate frame rate of 3.3 fps with the above-mentioned camera and the custom-made ultra-short-duration flash with a blue filter until the tracer was homogenously distributed, it was possible to determine θ. The colour change was also recorded in video formats with continuous lighting (in this case, there was no caption of the bubbles). The frames of the video were taken apart with the software Kinovea (Kinovea 0.8.15, Kinovea, France), and the same method used for analyzing the images was employed.

The mixing time θ was based on the time profile of the variation of the average colour pixel intensity (Rodriguez et al., 2013) on the vertical surface of the *µBC*. The acquired pictures and frames were transformed into *RGB* format with ImageJ (ImageJ 1.49v, National 211 Institutes of Health, Bethesda, Maryland, USA), and therefore the average colour pixel intensity was obtained as a combination of the intensities of the red, green and blue channels (example in Fig. 3–4).

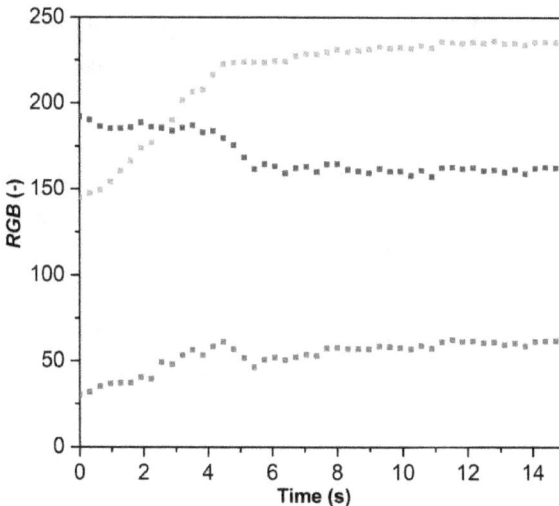

Fig. 3–4 Change of the *RGB* channel average values over time after the insertion of the fluorescent tracer in the glass-based microbubble column-bioreactor (*gµBC*) at a superficial gas velocity (u_G) of $1.3 \cdot 10^{-3}$ m/s (Lladó Maldonado et al. 2018).

The red and blue channel outputs varied the least (~30), while the green channel showed a maximum variation of ~ 90. Therefore, the green channel was selected to measure the mixing time. The time evolution of the green channel was characterized by a sigmoidal shape, which is consistent with the colour change from the dark blue background to green induced by the tracer insertion. The blue channel experienced a decrease in intensity, consistent with the background behind the transparent water, becoming green due to the fluorescein insertion. The red channel showed only a small increase due to the yellowish touch of the fluorescein (yellow is actually the combination of red and green). The green channel data was normalized for every mixing time experiment as described in eq. (3-8):

$$G^*(t) = \frac{G(t)-G(t_0)}{G(t_\infty)-G(t_0)} \qquad (3-8)$$

where $G(t)$, $G(t_0)$ and $G(t_\infty)$ denote the average pixel intensity of the vertical surface of the μBC at time t, at the start (t_0) and at the end of the experiment (t_∞), respectively. The liquid phase inside the μBC was considered to be homogenous when $G^*(t)$ was 0.95. The time at which this criterion is met was defined as the mixing time θ_{95}. The mixing time experiments were compared with CFD simulations (chapter 3.2.2) for a deeper understanding of the fluid dynamics.

Oxygen transfer

The volumetric liquid-phase mass transfer coefficient ($k_L a$) was measured by the dynamic gassing-out method. For every $k_L a$ measurement, oxygen was removed from the cell-free cultivation media by gassing nitrogen into the μBC until the oxygen concentration was equal to zero. Then, the nitrogen supply was turned off, and the reactor was aerated. $k_L a$ values were calculated from the increase in DO over time. The air flow was determined using a custom-made U-tube manometer as reported in Poterat et al (Peterat et al., 2014). The $k_L a$ values depended on the air flow supplied.

Fluid dynamics

The image series were analyzed using the software ImageJ (ImageJ 1.49v, National Institutes of Health, Bethesda, Maryland, USA) to determine the bubble size. The region of interest was selected through the cropping borders tool and was transformed into a binary image. The scale was set manually, taking the width of the reactor chamber as a reference. When measuring the bubble diameter, to select single gas bubbles and exclude dense bubble clusters, the circularity thresholds level was adjusted at 0.8 to 1.0.

The same image series from the bubble size analysis were used to calculate ε_G. In this case, the volume occupied by the aerated liquid at every u_G was compared to the volume liquid of the non-aerated system.

The bubble's movement was recorded with a high-speed camera (Exilim EX-FH20, Casio Europe GmbH, Norderstedt, Germany), which recorded at a frame rate of 290 fps. Through frame analysis of the videos, the bubble rise velocity (u_b) was calculated.

The local Reynolds number was calculated by eq. (2-10) applying the liquid velocity (υ_L), instead of the superficial liquid velocity (u_L).

3.2.2 Computational fluid dynamics simulations

CFD was employed as a complement to this research, acting as a supporting numerical tool to estimate the fluid behaviour inside the *gµBC*. Two-phase simulations were performed to model the three-dimensional fluid velocity field of both phases at different airflow rates. To reduce the number of mesh elements and computational time, only a half volume of the reactor was simulated by introducing a symmetry boundary condition in the xy-plane in the center of the reactor ($z=0$, where the nozzle was placed) (where the x-axis, y-axis and z-axis represent the direction of the width, height, and depth of the reactor, respectively), maintaining in this manner the three-dimensional geometry, which has been reported to yield closer predictions to the experimental data in bubble columns (Ekambara et al., 2005; Jakobsen et al., 2005; Krishna and van Baten, 2001). The symmetry condition is applicable since the flow behind the plane is a mirror of the other half. The first step towards the *CFD* simulation was to design the geometry of the body and discretize it into finite elements with a 3D *CAD* program (SolidWorks, Dassault Systèmes S.A, Vélizy, France). The second step was to generate a mesh with ANSYS ICEM CFD 17 (ANSYS Inc., Canonsburg, Pennsylvania, USA). A structured mesh approach (with hexahedral mesh elements) was preferred because it is well suited for flows with a dominant direction, the vertical flow in this case, and yields higher accuracy and better convergence than unstructured meshes, although they require longer preparation time. Three different structured mesh sizes were investigated, as well as the influence on the quality of the results: coarse, medium and fine meshes with 663, 1809, and 6195 mesh elements per µL, respectively. The finest mesh was selected because it offered the best correlation with respect to the steady-state and convergence considerations.

Steady-state computational fluid dynamics simulations

Bubble columns are well known to be highly transient systems, but as a considerable simplification approach in this work, steady-state simulations were performed. Two-phase steady-state *CFD* simulations were executed with the software ANSYS CFX 17 and 18 (ANSYS Inc., Canonsburg, Pennsylvania, USA), using a simplified model based on an Eulerian approach that solved the fluid governing equations, such as the Navier-Stokes and the continuity equations. The liquid fluid was defined as a continuous liquid phase, while the gas phase was implemented as a dispersed gas phase. A constant bubble size was introduced using the d_{VS} determined previously from the image analysis. The equations were solved without the use of turbulent models since the flow in microsystems is well known to be laminar, and proved in 3.3 with the calculation of Re_b yielding values of 1.4 – 6.9. The earth gravity field was included in order to consider the influence of buoyancy. The non-free surface model was applied for the interphase momentum, together with a surface tension coefficient between water and air of 0.072 N/m and a drag coefficient of 0.44. Regarding the boundary conditions of the model, the walls of the *gµBC* were modeled with a non-slip condition (zero velocity at the wall), the symmetry as a free-slip wall (no friction between fluid and wall), the nozzle as an inlet (constant gas flow velocities normal to the boundary condition were given), and the top surface of the *gµBC* as an opening (gauge pressure set to 0). Different steady-state simulations were carried out employing the same gas flow rates tested experimentally. The time step and convergence criteria were set to a maximal number of iterations of 200, an auto timescale with a conservative length scale and a residual acceptable value target Root Mean Square (*RMS*) of 10^{-4}, which is typically used.

Transient computational fluid dynamics simulations

Transient *CFD* simulations were carried out based on the previous converged steady-state simulations for the determination of the tracer distribution through the liquid volume of the aerated reactor simulating the mixing time experiments. The flow fields obtained from the steady-state simulations were frozen and used during the mixing study investigation. The tracer was simulated as an additional scalar variable with the diffusion coefficient of fluorescein, $4.25 \cdot 10^{-10}$ m^2/s (Kapusta, 2010), with the assumption that this component is dissolved in the water phase and transported accordingly along the velocity field and diffusion. The introduction of the tracer was carried out using a step function expression, filling with tracer the region next to the liquid inlet to emulate the step injection of the fluorescein applied in the mixing time experiments. The time step was set to 0.001 s, and the convergence control was set to a maximum of 3 coefficient loops and a residual target *RMS* of less than 10^{-4}. These conditions accomplish the Courant number criterion (number of passed mesh elements in each iteration step < 1), yielding a maximum Courant number of $5.5 \cdot 10^{-4}$.

Due to the simplifications in the model, it was expected that a significant deviation would occur between the simulation results and the experimental results. The interest in this numerical study was therefore to evaluate if the prediction quality was sufficient for the purposes of this work, which means that an approximate time scale for the mixing time was expected rather than a quantitative precise prediction.

3.2.3 Cultivation of *Saccharomyces cerevisiae*

A batch cultivation for validation purposes was performed with *Saccharomyces cerevisiae* strain CCOS 538 at 30 °C with a chemically defined cultivation medium as previously reported (Edlich et al., 2010; Peterat et al., 2014), *pH=* 4.5, with an initial glucose concentration of 20 g/L and an aeration with a u_G of $0.9 \cdot 10^{-3}$ m/s and $6.4 \cdot 10^{-4}$ m/s in the *gμBC* and in the *cμBC*, respectively. The air supplied through the nozzle was previously water-saturated. It was conducted through a bottle (500 mL) filled with 50 – 75 % distilled water set to a temperature of 30 °C to guarantee no cooling effect of the air on the liquid reaction and no evaporation effect.

Inocula of *S. cerevisiae* CCOS 538 were prepared from a cryo-culture stored with glycerol at −80 °C. The cells were reactivated by growing them overnight in round-bottom, sterile plastic cell culture tubes at 30 °C and shaking at 180 1/min, and the cultures were then diluted to an optical density measured at $\lambda = 600$ nm OD_{photo} of 0.3 (Spectrophotometer Libra S 11, Biochrom, Cambridge, UK) using deionized water as the reference. To measure OD_{photo} values within the linear range ($0.1 \leq OD \leq 0.45$), the samples were diluted using deionized water when necessary.

The experiments were performed in a custom-made incubation chamber (450 mm × 750 mm × 450 mm) that had temperature control and a heating and cooling system, as described in Peterat et al. (Peterat et al., 2014) and Krull and Peterat (Krull and Peterat, 2016).

3.2.4 Cultivation of *Staphylococcus carnosus*

The cultivations of *S. carnosus* were performed in a complex media described previosuly (Davies et al., 2013), with 1 g/L glucose and 10 µg/mL chloramphenicol as antibiotic, at 30 °C, with a pH value of 6.4 and with an aeration corresponding to a superficial gas velocity of $2.25 \cdot 10^{-3}$ m/s. Inocula of *S. carnosus* were prepared from cryo-cultures stored with glycerol at −80 °C. The cells were reactivated by growing overnight in shaken flask with 25 mL of the complex medium but with 10 g/L glucose at 30 °C and 180 1/min (25 mm eccentricity) and for starting the cultivations in the µBC, diluted to an OD_{photo} of 0.2 (Spectrophotometer Libra S 11, Biochrom, Cambridge, UK). The experiments were also performed in the custom-made incubation chamber.

For the continuous cultivation the feed flow was injected through the inlet of the µBC by using a precision syringe pump (neMESYS; Cetoni GmbH, Korbussen, Germany) in dispensing mode. The flow in the outlet was suctioned with also the precision syringe pump by using it in aspiration mode. The flows were adjusted according to the desired dilution rate by using the neMESYS UserInterface Software (Cetoni GmbH, Korbussen, Germany). The liquid handling was possible by using flexible Teflon tubing and cannulas (Sterican; B. Braun Melsungen AG, Melsungen, Germany).

For the analysis of the glucose concentration, the effluent was passed into interchangeable refrigerated sampling vessels. For rapid heat transfer, the sample vessels were cooled (Peltier element) in an aluminum block, which caused the samples to freeze as soon as they contacted the wall of the sampling vessels.

3.2.5 Heterogeneous biocatalysis

Immobilization procedure

During this study the non-covalent ionic immobilization of GOx to polyethylenimine-coated materials was conducted for reasons of simplicity. The enzyme immobilization procedure was carried out according to Mateo et al. (Mateo et al., 2000) with some changes. It involved the preparation of 1 mL immobilization mixture consisting of the corresponding amount of enzyme stock solution for a specific GOx loading filled up with 50 mM potassium phosphate buffer. This mixture was then added to 100 mg of polyethylenimine-modified, buffer-soaked particles and incubated for 3 h at room temperature in an end-over-end rotator at 20 1/min. The immobilization was then followed by four washing steps to remove unbound and weakly bound enzyme where the supernatant after short centrifugation was replaced with 1 mL phosphate buffer and kept on the end-over-end rotator for 5 min until the procedure was repeated.

Enzymatic activity measurement

The enzymatic activities of the GOx immobilizates were calculated from the depletion of the dissolved oxygen concentration during the oxidative biotransformation of β-D-glucose to D-glucono-1,5-lactone in two systems, a stirred glass beaker of 4 mL stirred at 300 1/min, referred here as miniaturised stirred tank reactor (mSTR), and in the microfluidised bed bioreactor (μFBR). The enzymatic activities in the μFBR, which had attached a microfluidic chip with an integrated glucose sensor (provided by the cooperation working group of Dr. Sesay, University of Oulu, Kajaani, Finland), were also determined using the depletion of glucose concentration. The oxygen depletion measurements were carried out in air-saturated 50 mM potassium phosphate buffer at 30 °C with 100 mM D-glucose. The linear part of the resulting slope was used for activity determination whereas one unit was defined as the consumption of 1 μmol oxygen/glucose per minute at the conditions mentioned before.

The observed activity A_{obs} calculations for the immobilizates in the mSTR were carried out according to eq. (3-9), and then divided by the weight of particles per mL:

$$A_{obs} \left(\frac{U}{mL} \right) = \frac{\frac{dc_L}{dt} \left(\frac{\mu mol}{min \cdot L} \right) \cdot 1/1000 \cdot \text{reaction volume } (mL)}{\text{sample volume } (mL)}$$

(3-9)

For calculation of the enzymatic activity inside the μFBR, it was necessary to take into account the oxygen transfer rate (OTR) provided by the stream of bubbles ($= k_L a \cdot (c_L^* - c_L)$). Eq. (2-5) allowed the determination of the OUR of the immobilized enzyme on the particles fluidised in the μFBR by insertion of the known $k_L a$, the saturation concentration of dissolved oxygen c_L^* and the current concentration of dissolved oxygen c_L.

The calculated OUR was then divided by the volume and concentration of added particle stock solution (0.1 g/mL) used inside the μFBR which resulted in the enzymatic activity in U/g. For activity determination with GOx loadings of 200 U/g or less, oxygen concentration depletion was not measurable since the oxygen uptake rate (OUR) was much lower than the OTR (except for MSU-F).

The activity calculation based on glucose concentration measurements inside the μFBR was carried out by firstly dividing the preliminarily measured 100 mM glucose peak signal by the obtained peak signals during the measurement. Afterwards, the determined polynomial regression function of the calibration (Fig. 3–6) was used to calculate glucose concentrations by converting the current ratios back to glucose concentrations.

The μFBR was tested in continuous mode with ReliZyme 403 S immobilizates with a GOx loading of 1000 U/g by running at a dilution rate of 13.09 1/h (flow rate of 2 µL/s) with a 100 mM glucose solution buffered with 50 mM KH_2PO_4 (pH 7). The same amount of medium that was pumped into the μFBR was removed by a second pump. The generated effluent at the outlet of the μFBR was collected in a glass vial placed inside a 70 °C thermomixer (MixMate, Eppendorf, Wesseling-Berzdorf, Germany) for 15 min to ensure the deactivation of potentially leaked enzyme. Every five minutes, the glass vial was exchanged and the concentrations of glucose and gluconic acid were determined by HPLC as explained in chapter 3.2.6.

3.2.6 Sensors

Optical density

The OD of the biomass, which was measured at 600 nm ($OD_{\mu BC}$) during cultivation, was determined online using an LED panel (EA LG40X21-A green-yellow, 51 x 21.2 x 4.8 mm, 8 V, Electronic Assembly, Gilching, Germany) and a miniature spectrometer (USB 2000+, Ocean Optics, Ostfildern, Germany) was coupled to an optical fiber (200 µm diameter, M24L05, Thorlabs, Dachau, Germany). The $OD_{\mu BC}$ data ($OD_{\mu BC} = ln\,(I_0/I)$, where I_0 and I are the light intensity measured through the cell free cultivation medium and the cell suspension, respectively) was continuously measured every second and an average of ten monitoring points was recorded every 10 s.

A correlation between $OD_{\mu BC}$ and optical density measured offline in the spectrophotometer (OD_{photo}) needed to be determined for every cultivation. A linear correlation was adjusted with three pairs of OD measurements data (media, inocula and final) and that enabled the conversion of $OD_{\mu BC}$ to OD_{photo}.

The correlation between the OD_{photo} and biomass concentration, determined as the cell dry weight (CDW) concentration here referred as c_{CDW} (g/L) of S. cerevisiae, was derived from measurements of samples with known CDW concentrations, as reported previously by Peterat (Peterat, 2014) in eq. (3-10):

$$c_{CDW} = 0.3743 \cdot OD_{photo} \qquad\qquad (3\text{-}10)$$

The correlation of S. carnosus, was derived from measurements of samples with known biomass concentration, given by eq. (3-11):

$$c_{CDW} = 0.0501 \cdot OD_{photo} \qquad\qquad (3\text{-}11)$$

The CDW concentration was determined by filling 10 mL of the cultivation broth in a test tube, and centrifuged it at 3000 1/min for 10 min. The supernatant was removed and afterwards the pellet was resuspended in 10 mL 0.9 % sodium chloride solution to remove medium residues. A second step of centrifugation was done, and then the pellet was dried at 105 °C until constant weight.

Dissolved oxygen

In the case of the glass μBC, dissolved oxygen measurements were carried out using a needle-type oxygen microsensor (OXR50-CL4, Pyroscience, Aachen, Germany) connected to a four-channel phase-shift fluorimeter (Firesting, Pyroscience, Aachen, Germany).

DO concentration measurements in the $c\mu BC$ were carried out by integrating oxygen sensor spots on the inside of the microscope slide that closed the vertical back side of the $c\mu BC$ in cooperation with Prof. T. Mayr, Institute of Analytical Chemistry and Food Chemistry, Graz University of Technology, Austria. First, a primary layer of the polymer HydroMed D4 hydrogel (AdvanSource Biomaterials Corp., Wilmington, Massachusetts, USA) dissolved in a mixture of tetrahydrofuran (THF) and toluene was microdispensed onto the microscope slide to improve the adhesion of the DO sensor spots. The oxygen sensor cocktail was prepared by dispersing oxygen sensitive particles in D4 hydrogel dissolved in isopropanol and water mixture. The particles were prepared according to Nacht et al. (Nacht et al., 2015) by polymerization and afterward stained with an oxygen sensitive dye, platinum(II) meso-tetra(4-fluorophenyl)tetrabenzoporphyrin (PtTPTBPF).

pH sensor

pH sensors were also prepared in cooperation with the group of Prof. T. Mayr, Institute of Analytical Chemistry and Food Chemistry, Graz University of Technology, Austria. pH sensor spots were micordispensed onto the microscope slide of the μBC. Similar to the oxygen sensor spots, a primary layer of polymer HydroMed D4 hydrogel dissolved in toluene was first spread on the polymeric slide to improve the adhesion of the sensor spots. The pH sensor cocktail was prepared by dissolving the pH-indicator aza-BODIPY in D4 hydrogel in THF and water and adding Egyptian Blue reference particles, subsequently. pH sensitive aza-BODIPY dye (4-(7-(3,5-dichloro-4-hydroxyphenyl)-5,5-difluoro-1,9-diphenyl-5H-5λ^4,6λ^4-dipyrrolo[1,2-c:2',1'f][1,3,5,2] triazaborinin-3-yl)-N-dodecylbenzamid) was synthesized in-house according to Strobl et al. (Strobl et al., 2015). Egyptian blue reference particles ($CaCuSi_4O_{10}$) were produced according to Borisov et al. (Borisov et al., 2013). The composition of the sensor cocktails as well as that of the primary layer used for microdispensing is described in Tab. 3–2.

Tab. 3–2 Composition of the sensor cocktails and primary surface modification layer used for microdispensing the chemical sensors. Adapted from Sun, 2017.

Cocktail	D4 (mg)	THF (mg)	Toluene (mg)	Isopropanol (mg)	Water (mg)	Particles/ Indicator (mg)	Egyptian Blue (mg)	Total (mg)
Primary layer – D4	250	2012.5	862.5	N/A	N/A	N/A	N/A	3125
Oxygen layer in D4	250	N/A	N/A	2937.5	979.17	250	N/A	4416.67
pH layer in D4	500	5175	N/A	N/A	575	1.5	250	6501.5

The microdispenser (MDS 3200A, VERMES Microdispensing GmbH, Otterfing, Germany) was mounted on a three-axis Computerized Numerical Control machine (assembled in-house) run by the freeware LinuxCNC. The guiding programs for dispensing were written in G-Code. The microdispensing device worked with a piezoelectrically guided tappet that "shot" fluid drops (in this case polymer solutions) through a nozzle. The cocktail reservoir was sealed from the outside and under pressure. Adequate layer thickness was verified by checking the signal intensity of the spot.

Oxygen and pH sensor spots were read-out with a four-channel phase-shift fluorimeter FireStingO2 (PyroScience, Aachen, Germany) and a one-channel phase-shift fluorimeter Piccolo2 (PyroScience, Aachen, Germany). These instruments were used for luminescence lifetime-based measurements in the frequency domain. Dual lifetime referencing was applied for pH sensors as reported by Jokic et al. (Jokic et al., 2012) and Borisov et al. (Borisov et al., 2013). A custom-made holder made of polycarbonate was used to hold and fix these fibers at their same positions throughout the experiments. The optical fibers (tip diameter 1 mm) were directed to the outer surface of the µBC, where they were guided through access holes in the chip holder and connected to FireStingO2 for the online oxygen monitoring spot and to Piccolo2 for the online pH monitoring spot. Both the DO and pH data were continuously measured and recorded every 10 s from an average of three measurements using Pyro Oxygen Logger software (PyroScience, Aachen, Germany).

Calibration data for the oxygen sensor spot were similar to reported in Nacht et al. (Nacht et al., 2015). The oxygen sensors were calibrated via a two-point calibration. The μBC was filled with cultivation media and then flushed with air and N_2, respectively to determine the phase shift (*dphi*) at air-saturated and deoxygenated conditions. To calibrate the *pH* sensor spot, 5 solutions of cultivation media were adjusted to different *pH* values (*pH*= 1.85, 3.25, 3.95, 4.71 and 6.39) by adding *HCl* or *NaOH*, the *pH* of which were measured by a *pH* electrode (BlueLine 16 pH, SI Analytics, Weilheim, Germany) and a *pH* meter (CG840, Schott-Geräte, Mainz, Germany). Each of these calibration solutions were used to fill the *cµBC*, and the phase shifts were recorded. The calibration environment (e.g., temperature, aeration, and light) was the same as that of the cultivation conditions, and the calibrations were performed by measuring the *dphi* values. A Boltzmann curve was fitted to the calibration points (cotangent of the *dphi* values with respect to the *pH*) with OriginPro 2015 (OriginLab Corporation, Northampton, Massachusetts, USA) (Fig. 3–5). The dynamic range of the sensor was from ~ 2.5 to 5.5.

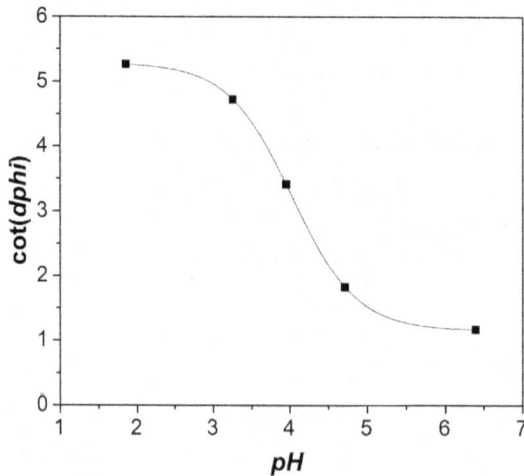

Fig. 3–5 Calibration of the *pH* sensor in the cuvette-based microbubble column-bioreactor (*cµBC*). To obtain the calibration curve, the data (cotangent of the recorded *dphi* values with respect to the *pH*) were fitted to a Boltzmann curve (Lladó Maldonado et al. 2019a).

Glucose sensor

The substrate-limiting carbon source glucose consumed by the cultivated cells, normally requires sampling and bacterial removal to measure its concentration offline. However, here, it was measured online using an electrochemical biosensor integrated in an adjacent microfluidic chip as sampling was not possible because of the small $c\mu BC$ volume.

Glucose depletion was monitored by a glucose biosensor based on a glucose oxidase enzymatic electrochemical biosensor, as reported in Panjan et al. (Panjan et al., 2017b). The glucose sensor was established in cooperation with the group of Dr. Sesay, Unit of Measurement Technologies, University of Oulu, Kajaani, Finland. The sensor consisted of screen-printed electrodes (3 mm diameter graphite working electrode, graphite counter electrodes and sliver reference electrode) (ECOBIO lab, Florence, Italy) modified with a layer of immobilized GOx from Aspergillus niger VII S (195 U/mg) (E.C. 1.1.3.4) as the catalytic biological recognition element. Prussian blue (PB) was used to modify the surface of the working electrode to perform as a redox mediator. The hydrogen peroxide produced by GOx caused a selective electrocatalytic reaction with PB. The current was measured by amperometric detection using a potentiostat (PalmSens, Houten, The Netherlands). The amperometric signal which directly correlates to the concentration of glucose in the bulk solution was used for the quantification of glucose. Amperometric measurements were calibrated via known concentrations of glucose in the cultivation broth (at the beginning and at the end of the cultivation).

The glucose biosensor was integrated into a PMMA microfluidic chip that was fabricated using a CO_2-laser and bonded using double-sided medical grade 100 μm thick adhesive (Adhesive Research, Limerick, Ireland). The biosensor integration chip featured a microfluidic loop that allowed an internal buffer (phosphate buffer saline, PBS) to flow using a precision syringe pump (Nemesys, Cetoni GmbH, Korbussen, Germany) between the integrated semipermeable membrane (contact area 12 mm^2, cut-off molecular weight 6 - 8 kDa, Spectrum Laboratories Inc., Rancho Dominguez, California, USA) and the downstream integrated glucose biosensor. The microfluidic flow chip was mounted onto the backside of the microscope slide and integrated into the *cµBC* via a drilled hole (3.5 mm in diameter), such that the semipermeable membrane was in contact with the cultivation broth. Only molecules as small as glucose could diffuse through the membrane. The microfluidic flow chip extended the concentration range of glucose detection by working as a dilution unit of the bulk solution, depending on the chosen diffusion time. The glucose biosensor had a limit of detection of 0.006 mM (1 mg/L), with a linear range of up to 3 mM (5.4 mg/L) (Panjan et al., 2017b).

A stopped-flow protocol was used for the measurement of glucose. The integrated microfluidic flow chip buffer flow was stopped for 7.5 min (diffusion time) to allow glucose to diffuse from the *MBR* into the buffer underneath the semipermeable membrane. After the diffusion time, the flow was turned on again for 1.5 min (flow rate 1.67 µL/s), and the glucose concentration in the buffer could flow to the other side of the membrane and be measured when it was in contact with the glucose sensor.

Since *GOx* that was used for the enzymatic studies inside the *µFBR* produced hydrogen peroxide, too, the measured values for the glucose concentrations were affected by the diffusing hydrogen peroxide. To characterize the extent of this interference, a calibration was conducted at the end of each set of loaded carrier materials. Therefore, a constant signal for 100 mM glucose was ensured which was followed by measurement of expected *GOx* reaction mixtures consisting of final glucose, gluconic acid and hydrogen peroxide concentrations inside the *µFBR* to mimic expected biotransformation concentrations. This was carried out in a glucose concentration range between 99 and 95 mM and 1 up to 5 mM for gluconic acid and hydrogen peroxide, respectively. Fig. 3–6 shows two independent glucose sensor calibrations which were performed with two different biosensors and thus varying signal intensities. Despite these differing starting conditions, the relation between the 100 mM glucose peak and the following expected glucose oxidase mixtures gave a constant correlation.

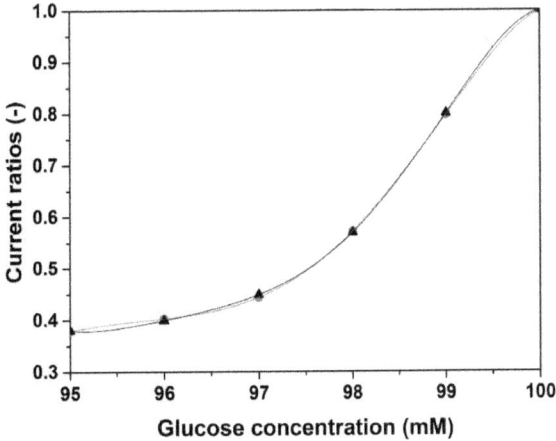

Fig. 3–6 Current ratios and polynomial regression functions from glucose sensor calibration. ▲ and ● indicate two independent calibrations with different glucose biosensors. Determined polynomial regression function for ● was: $y = -6.253 \cdot 10^{-4} \ x^5 + 0.301 \cdot x^4 - 57.899 \cdot x^3 + 5569.574 \cdot x^2 - 267822.480 \cdot x + 5150330$ with $R^2 = 1$.

The activity calculation based on glucose concentration measurements inside the µFBR was carried out by firstly dividing the preliminarily measured 100 mM glucose peak signal by the obtained peak signals during the measurement. Afterwards, the previously determined polynomial regression function of the calibration (Fig. 3–6) was used to calculate glucose concentrations by converting the current ratios back to glucose concentrations.

Glucose offline measurement

For the continuous experiments it was possible to have glucose offline measurement of the samples collected in the outlet. For the continuous cultivation of S. carnosus the measurement was performed with the glucose analyser Kreienbaum YSI 2900/2950 (YSI Incorporated, Yellow Springs, Ohio, USA). On the other side, for the continuous enzymatic experiments the glucose and gluconic acid concentrations were determined via HPLC according to Walisko et al. (Walisko et al., 2017) and Adler et al. (Adler et al., 2014), respectively. The quantification of gluconic acid was slightly modified; it was performed via UV detection at 210 nm.

4 Characterization of a multiphase microbioreactor

The aim of this chapter was to present the fluid dynamic and mass transfer characterization of the borosilicate glass-based microbubble column-bioreactor ($g\mu BC$) and show the influence of superficial gas velocity (u_G) on oxygen transfer rate (*OTR*) and mixing time θ. The objective was to prove that the pneumatic aeration through a micro nozzle provided sufficient aeration to satisfy the high demand of oxygen of aerobic bioprocesses, and at the same time, it enabled the homogenization of the cultivation broth. This investigation offered a further improvement on the $g\mu BC$ based on a deeper engineering characterization. Parameters such as Sauter mean diameter of bubbles (d_{vs}), gas hold-up (ε_G), mixing time (θ), or bubble rise and liquid velocity (u_b and u_L, respectively) could be obtained. Additionally, a simplified *CFD* model was developed as a complement to this research as a supporting numerical tool to estimate the fluid behaviour inside the $g\mu BC$.

4.1 Fluid dynamics

The stream of air bubbles plays an important role in the $g\mu BC$. It avoids cell sedimentation, provides a good cultivation broth mixing, and induces the oxygen transfer from the gas into the liquid phase. For this reason, a large effort was made to characterize the fluid dynamics of the $g\mu BC$.

4.1.1 Airflow rate

The resulting Q_G and u_G were determined with respect to the applied air pressure into the nozzle of the μBC. The flows were measured in triplicates, and the mean results together with the respective standard deviation are shown in Fig. 4–1. For air pressures from 1 to 3.5 bar, Q_G ranged from 0.6 to 6.2 µL/s, and u_G respectively ranged from $0.2 \cdot 10^{-3}$ to $2.2 \cdot 10^{-3}$ m/s. As expected, at higher air pressures, higher flow rates were provided. The gas volume flow rate increased reasonably linearly with the pressure. Deviations from the linearity may be caused by the change of the gas flow resistance.

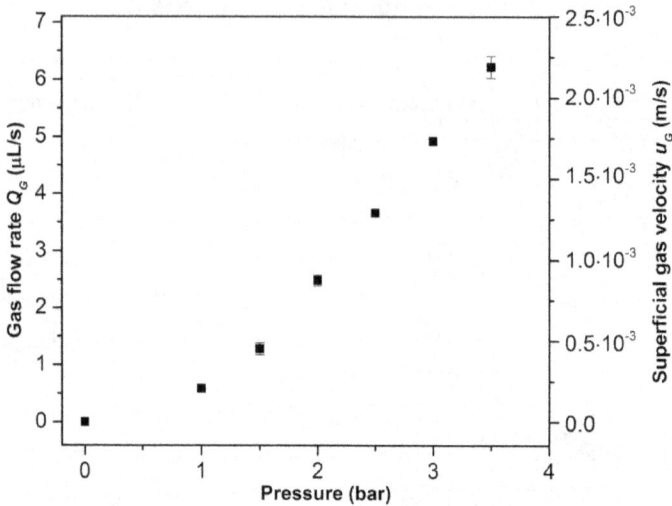

Fig. 4–1 Relation between the supplied air pressure through the nozzle of the glass-based microbubble column-bioreactor (*gμBC*) and the resulting gas flow rate (Q_G) and superficial gas velocity (u_G) calculated by eq. (3-7). Each point represents the average of triplicate flow measurements. Error bars show standard deviations (Lladó Maldonado et al. 2018).

4.1.2 Bubble characterization and gas hold-up

By taking single images with an ultra-short-duration flash, it was possible to observe the gas flow pattern and the bubble distribution for different gas velocities (Fig. 4–2). For the u_G range between $0.2 \cdot 10^{-3}$ and $0.5 \cdot 10^{-3}$ m/s, it was easy to distinguish a column of bubbles discharged from the nozzle, indicating a suboptimal mixing regime. Because u_G exceeded $0.9 \cdot 10^{-3}$ m/s, there was a good distribution of the bubbles in the *gμBC*, being defined as a homogenous bubbly regime (Bouaifi et al., 2001). With increasing u_G, the bubbles started to rise in clusters, but always inside the homogenous bubbly regime mentioned above.

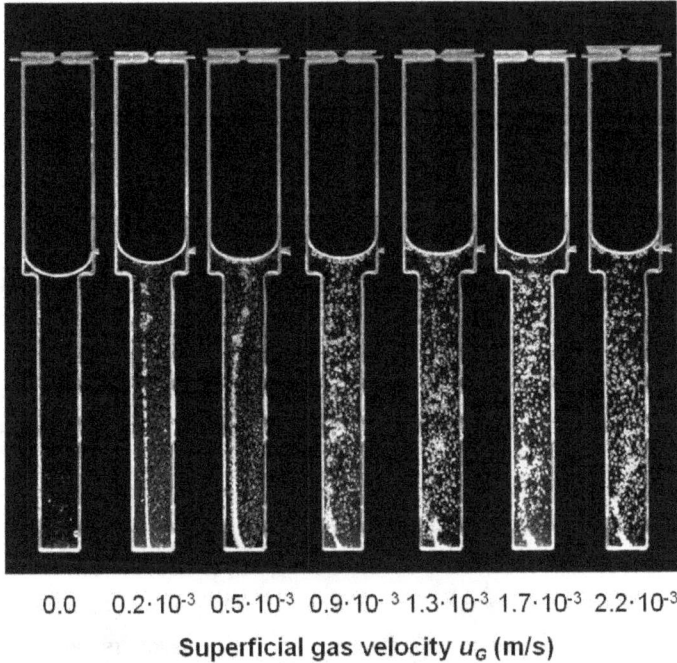

0.0   0.2·10⁻³ 0.5·10⁻³ 0.9·10⁻³ 1.3·10⁻³ 1.7·10⁻³ 2.2·10⁻³

Superficial gas velocity u_G (m/s)

Fig. 4–2 Effect of the airflow rate (Q_G) and superficial gas velocity (u_G) on the gas flow pattern of the glass-based microbubble column-bioreactor ($g\mu BC$) (Lladó Maldonado et al. 2018).

The same pictures were analyzed to calculate d_{vs} and ε_G for every u_G. d_{vs} was analyzed based on three images at every u_G (Fig. 4–3 A) ranging from 50 to 150 µm. Even smaller bubbles could be observed, but they were not measurable due to photography limitations. The size of the bubbles increased with increasing u_G, because larger bubbles are being generated at the sparger. In this type of regime, there is practically no bubble coalescence or break-up (Kantarci et al., 2005). A similar increasing tendency in ε_G was observed as u_G was increased (Fig. 4–3 B), taking values between 0.06 and 0.09.

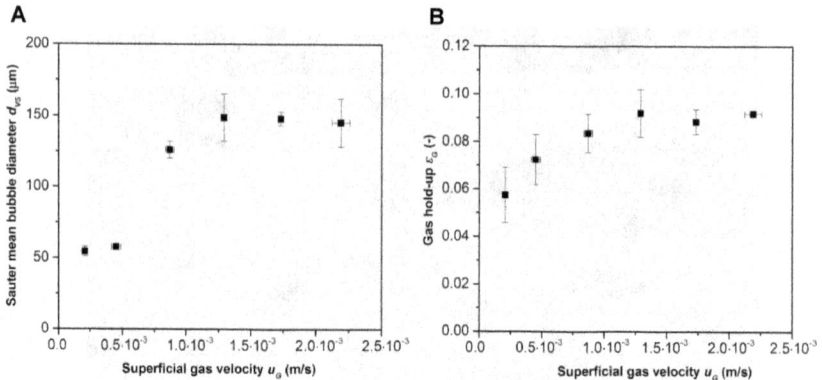

Fig. 4–3 Effect of the superficial gas velocity (u_G) on (A) the Sauter mean bubble diameter (d_{vs}) calculated by eq. (2-6) and (B) the gas hold-up (ε_G). Each point represents the average of triplicate measurements. The vertical error bars represent the standard deviation of the triplicate measurements of d_{vs} and ε_G, respectively, and the horizontal error bars represent the standard deviation of the averaged u_G determined in chapter 4.1.1. Adapted from Lladó Maldonado et al. 2018.

Peterat et al. (Peterat et al., 2014) obtained similar ε_G values with the same nozzle size as the current μBC. Regarding the size of the air bubbles, it was observed that they are smaller in the current μBC, meaning that the formation frequency is higher for the same u_G. This is translated into more interfacial area between the gas and liquid phases, which improves the oxygen transfer. When working with μBCs, it is important to match the bubble size range between 50 and 150 µm, as otherwise, when the bubbles are too small, foam could occur, or with bigger bubbles, the liquid could be pumped out of the reactor.

4.1.3 Reynolds number of the rising bubbles and volumetric power input

The bubble rise velocity (u_b) remained relatively uniform, which is typical for homogenous bubble flows (Weuster-Botz et al., 2001), at 0.025 m/s but experienced an increase at the higher u_G values tested. From the u_b values together with the d_{vs} values, it was possible to calculate the Reynolds number of the rising air bubbles (Re_b) (Fig. 4–4 A). Re_b increased with increasing u_G, ranging from for the tested u_G spectrum. These values indicated laminar flow conditions. In Doig et al. (Doig et al., 2005b), Re_b reached values up to 196 higher than in the $g\mu BC$, because of the larger size of the bubbles.

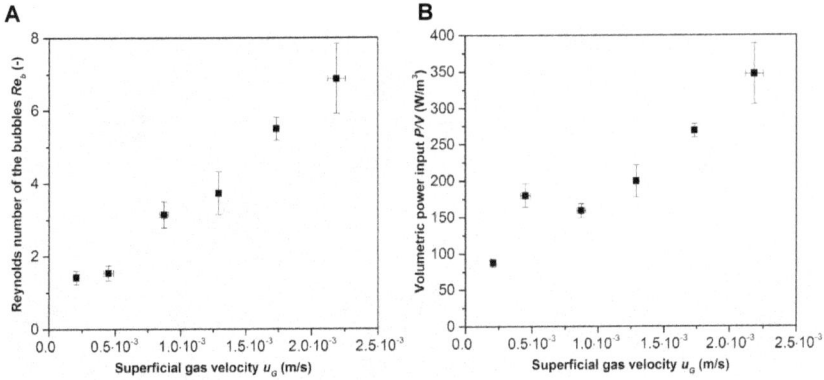

Fig. 4–4 Effect of the superficial gas velocity (u_G) on (A) the Reynolds number of the bubbles (Re_b) calculated by eq.(2-8) and (B) the volumetric power input (P/V) calculated by eq.(2-9). Each point represents the average of triplicate measurements. The vertical error bars represent the standard deviation of the triplicate measurements of Re_b and P/V, respectively, and the horizontal error bars represent the standard deviation of the averaged u_G determined in chapter 4.1.1. Adapted from Lladó Maldonado et al. 2018.

Power input due to the expansion work of the gas bubble is neglected due to the small height of the fluid column in the $g\mu BC$. The opposite occurs for bubble columns, where the power input caused by gas bubbling at the sparger can be neglected in comparison to the expansion work due to the large height of the column (Weuster-Botz et al., 2001). The volumetric power input (P/V) caused by bubble generation at the nozzle versus the superficial gas velocity (u_G) is depicted in Fig. 4–4 B. P/V tends to increase with rising u_G, reaching values up to 350 W/m^3. These values are of the same magnitude as in miniaturised bubble column reactors; for instance, Weuster-Botz et al. (Weuster-Botz et al., 2001) reached P/V values up to 100 W/m^3 in a 200 mL-bubble column reactor. Each of these are one order of magnitude smaller than that determined for the pilot or large-scale bubble columns (the order of magnitude of 10^3 W/m^3) (McClure et al., 2014).

4.1.4 Steady-state computational fluid dynamics simulations, superficial liquid velocity, and the Reynolds number of the liquid

Two-phase simulations were performed to model the fluid velocity and fluid velocity field at different airflow rates. The spatial distribution of the liquid velocity (v_L) and the local Reynolds number of the liquid (Re_L) (calculated by eq. (2-10), applying v_L instead of u_L) obtained from the simulation are presented in Fig. 4–5.

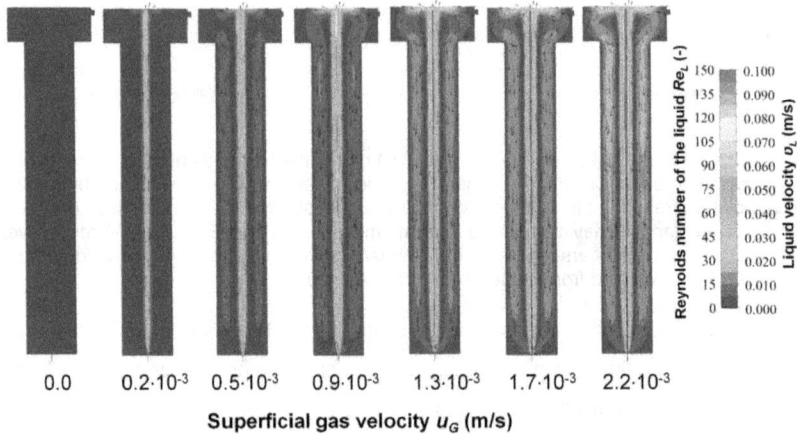

Fig. 4–5 Two-phase simulations to model the liquid velocity (v_L) and Reynolds number of the liquid (Re_L) fields at different superficial gas velocities (u_G) in the glass-based microbubble column-bioreactor ($g\mu BC$) (Lladó Maldonado et al. 2018).

The spatial distribution of v_L and Re_L are represented in a vertical cross-section through the center of the $g\mu BC$. The velocity vectors are also shown, so it is possible to see the recirculation motion of the liquid. The distribution of v_L changes dramatically depending on the coordinate in the xy-plane with respect to the nozzle. In the rising area, where the stream of bubbles is dragging the liquid upwards, the liquid is moving faster than in the downcomer section close to the walls of the reactor. In the downcomer section, the area is larger than in the riser section and there is the friction at the wall, reducing v_L. Re_L goes from values close to 0 next to the reactor walls up to ~150 for the areas in direct contact with the air bubbles. All the values in the range correspond to laminar flow conditions, typical in *MBRs*.

Fig. 4–6 presents the u_L and Re_L (calculated by eq. (2-10), applying u_L) in the entire volume estimated with the CFD model in the considered u_G regime. At higher u_G, higher liquid velocities were achieved, and more mixing was generated, increasing Re_L.

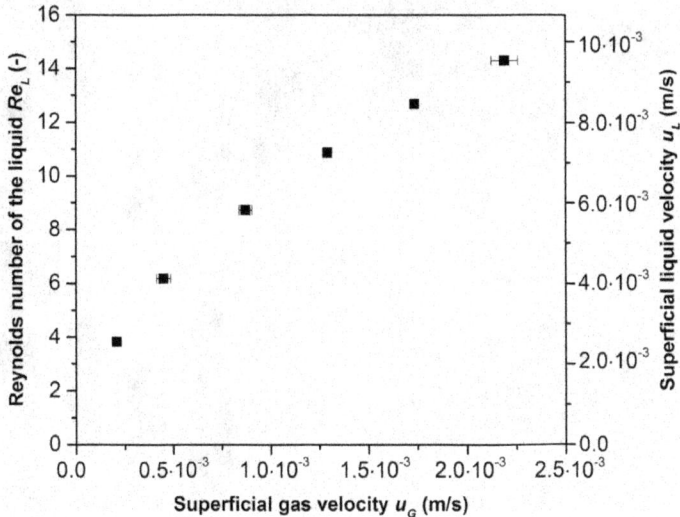

Fig. 4–6 Effect of the superficial gas velocity (u_G) on the numerically CFD estimated Reynolds number of the liquid (Re_L) calculated by eq. (2-10) and on the numerically CFD estimated superficial liquid velocity (u_L) (Lladó Maldonado et al. 2018).

4.2 Mixing time and transient computational fluid dynamics simulations

Mixing experiments were performed at different airflow rates. Fig 4-7 A shows an example of the sequence of images immediately after the injection of the fluorescent tracer pulse with a frame rate of ~1 fps at an aeration of u_G= 1.3·10^{-3} m/s. In addition to mixing laboratory experiments, simulated tracer profiles were calculated through transient CFD simulations. The tracer was introduced as a step function in the gμBC (1.6 % v/v). As an example, time-lapse image series of the transient simulation with the same aeration of u_G= 1.3·10^{-3} m/s are shown in Fig. 4–7 B with a frame rate of ~ 1 fps. When comparing the experimental and simulated tracer profiles, for the same aeration rate, the model is properly predicting the tracer profile distribution. This indicates that this model could be used to represent the oxygen, carbon substrate and cell dispersion.

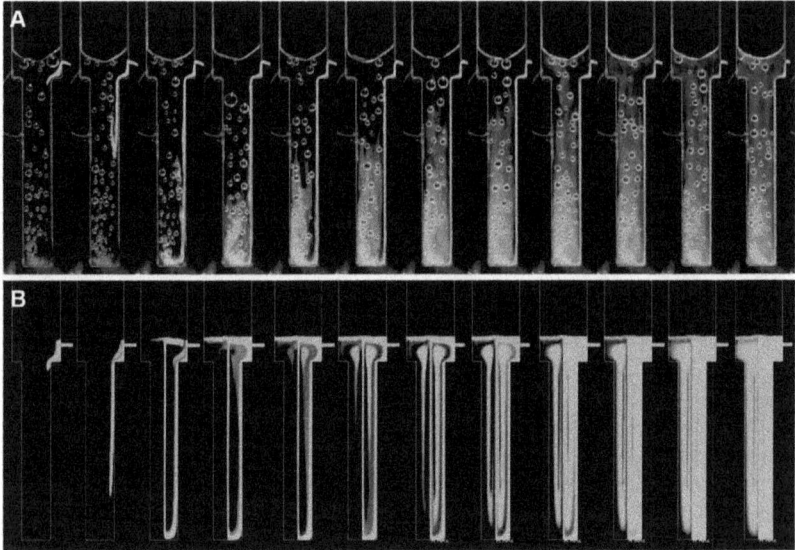

Fig. 4–7 Time-lapse image series of the glass-based microbubble column-bioreactor ($g\mu BC$) with a superficial gas velocity (u_G) set at $1.3 \cdot 10^{-3}$ m/s (A) after the experimental injection of a pulse of 2 µL of the fluorescent tracer solution through a needle pump and (B) the transient *CFD* simulation to model the mixing time experiments. The images are shown with a frame rate of ~ 1 fps (Lladó Maldonado et al. 2018).

Furthermore, all the images obtained from the laboratory experiments (average frame rate of 3 fps) and through simulations (average frame rate of 10 fps) were analyzed with ImageJ, applying the normalized green channel method. Fig. 4–8 presents an example of the variation in the normalized green-channel data $G*$ for the experimental mixing exercise (two replicas) and that obtained in the transient *CFD* simulations, both at a superficial gas velocity (u_G) of $1.3 \cdot 10^{-3}$ m/s. Although the predicted mixing time θ_{95} is reached later, at 11.8 s, compared with the 7.7 s obtained experimentally, the qualitative prediction is considered good, with an acceptable deviation. This could be because this model contains substantial simplifications and does not predict the dispersion of the bubbles in the entire cross-section, just a column of bubbles in the center of the $g\mu BC$. The mixing time deviation in the time scale of seconds becomes less significant when it is compared to the time scale of biological processes in the range of hours or days. The development of this simplified model will give the opportunity to analyze different reactor designs optimizing the geometry and flow conditions in further *MBR* design-manufacturing iterations, as well as acting as a starting reference for improving the model in future simulations.

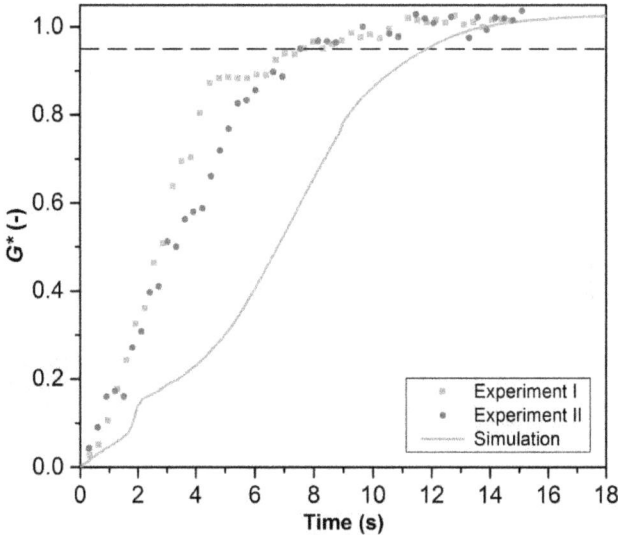

Fig. 4–8 Comparison of the measured and simulated variation of the normalized green-channel data G^* after the fluorescent tracer pulse in the glass-based microbubble column-bioreactor (*gμBC*), with a superficial gas velocity (u_G) of $1.3 \cdot 10^{-3}$ m/s, calculated by eq. (3-8) (Lladó Maldonado et al. 2018).

The mixing time θ_{95} was characterized for different u_G values (Fig. 4-9) ranging from 13 down to 5.5 s for higher airflow rates. θ_{95} ended up being inversely proportional to u_G. It was observed that θ_{95} experienced a minimum of 5.5 s, not decreasing further when u_G further increased. A similar procedure was followed in Kirk et al. (Kirk et al., 2016), processing images after the introduction of a dye into the *MBR*, resulting in mixing times of ~7 s. Peterat (Peterat, 2014) presented comparable values between 15 and 2 s in the u_G range of $0.12 \cdot 10^{-3}$ and $0.4 \cdot 10^{-3}$ m/s, determined through three photodiodes located at different positions. For the miniaturised 10 mL-stirred tank bioreactor, Betts et al. (Betts et al., 2006) determined a mixing time θ_{95} between approximately 20 and 5 s, depending on the agitation rate, to detect a 95 % *pH* shift.

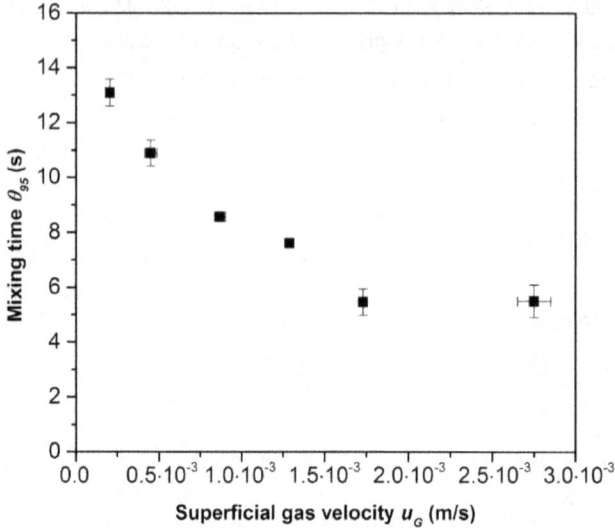

Fig. 4–9 Variation of the mixing time (θ_{95}) in the glass-based microbubble column-bioreactor ($g\mu BC$) in relation to the superficial gas velocity (u_G) applied. Each point represents the average of triplicate measurements. Error bars indicate standard deviations. Adapted from Lladó Maldonado et al. 2018.

4.3 Oxygen transfer

To evaluate possible applications, it was important to determine the maximal oxygen transfer rate (OTR_{max}) of the $g\mu BC$, which is often the limiting parameter in any bioprocess. The determination of $k_L a$ as a function of u_G was performed by interchanging the inlet gas flow between nitrogen and air. The average of the $k_L a$ values obtained from the slope of the oxygen concentration profiles (measured in triplicates) is shown in Fig. 4–10 together with the OTR_{max} values that are calculated by multiplying the respective $k_L a$ values and the concentration of dissolved oxygen (DO) in liquid at equilibrium with the gas phase concentration (c_L^*) at 30 °C, 7.56 mg O_2/L.

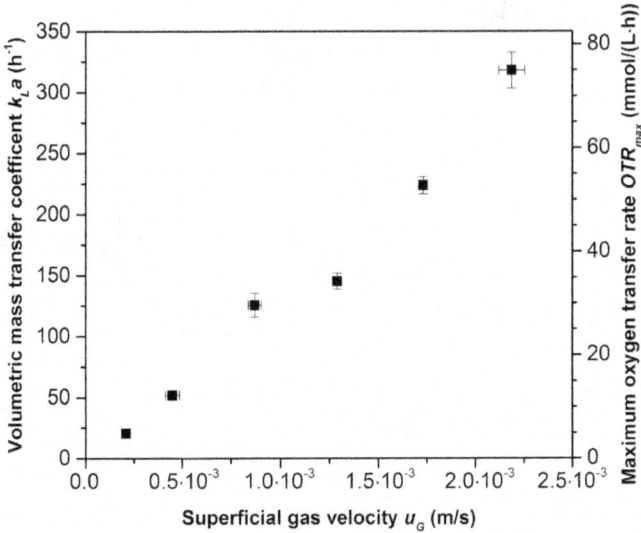

Fig. 4–10 Effect of the superficial gas velocity (u_G) on the volumetric liquid-phase mass transfer coefficient (k_La) calculated by eq. (2-2) and on the maximum oxygen transfer rate (OTR_{max}) calculated by eq. (2-3) for the glass-based microbubble column-bioreactor ($g\mu BC$). Each point represents the average of triplicate measurements. Error bars indicate standard deviations. Adapted from Lladó Maldonado et al. 2018.

The k_La values increased while increasing u_G, ranging from 20 to a maximum value of 320 1/h for the highest u_G tested. In the same context, the OTR_{max} varied from 5 to 75 mmol/(L·h). The k_La values measured in the present $g\mu BC$ were slightly higher than in a 250 mL-shaken flask filled with 10 % relative volume and shaken at a frequency of 180 1/min, with ~100 1/h (Maier et al., 2004). Peterat et al. (Peterat et al., 2014) estimated k_La values ranging from 72 to 400 1/h for the same u_G spectrum as in the present investigations and benefited from the gas permeability of the PDMS reactor wall, up to a maximum k_La value of 500 1/h.

The $g\mu BC$ could be compared to similar systems found in the literature but at milliliter scale, such as the 2 mL-miniature bubble column reactor from Doig et al. (Doig et al., 2005a; Doig et al., 2005b) that reached k_La values up to 220 1/h, the 20 mL-miniature bubble column from Kheradmandnia et al. (Kheradmandnia et al., 2015) with reported k_La values between 100 and 800 1/h, or the 200 mL-bubble column reactor from Weuster-Botz et al. (Weuster-Botz et al., 2001) that exceeded 540 1/h.

Other systems at microscale were recently reported, such as in Kirk et al. (Kirk et al., 2016), where the k_La of an oscillating jet-driven *MBR* with a working volume of 50 μL was found to reach values up to 170 1/h, or the 1 mL-bioreactor system of Bolic et al. (Bolic et al., 2016) with k_La values between 124 and 450 1/h in the sparging mode.

When Fig. 4–3 and Fig. 4–10 are compared, a relation between d_{vs} and ε_G with k_La values is perceptible. k_La is the product of the liquid-phase mass transfer coefficient (k_L) multiplied by the interfacial area per unit volume (a) between the gas and liquid phases. This specific phase boundary surface a depends on d_{vs} and ε_G. Higher values of a are obtained, and therefore higher k_La values, when d_{vs} is smaller, due to the higher surface-area-to-volume ratio that implies, and when ε_G is larger, which denotes more bubble content. It is noticeable that, at some point, despite d_{vs} and ε_G reaching a plateau (Fig. 4–3), the k_La values continue to linearly increase. This phenomenon can be explained by the fact that the momentum transfer increased with increasing u_G and that enhanced the internal liquid circulation, which is also in agreement with the numerical solution. Therefore, the absolute u_b increased and the mass transfer rate as well.

4.4 Cultivation of *Saccharomyces cerevisiae*

To validate the *gμBC* as a screening tool for biotechnological research, a batch experiment with *S. cerevisiae* was performed with the online monitoring of *DO* and biomass concentration. Fig. 4–11 shows agreement between the *DO* and *CDW* concentration during the cultivation time, where the *DO* slightly decreased as the *CDW* concentration increased. After a lag phase of 3 h the yeast cells grew exponentially with a specific growth rate of 0.29 1/h until the cultivation time of 10.5 h, before the growth slowed down, indicating the start of the diauxic phase. Compared to cultivation in a shaken flask, a very similar specific growth rate of 0.28 1/h was obtained. The inoculums did not seem to have an optimal viability. With the same strain and the same starting glucose concentration of 20 g/L Peterat et al. (Peterat et al. 2014) reached a shorter lag phase, a higher maximal specific growth rate of 0.37 1/h, and reached higher biomass concentrations. Despite the suboptimal conditions of the inoculum viability, the tendency of the growth curve fits well with the theoretical model of a batch cultivation. Regarding the oxygen transfer capacity of the *gµBC* the *DO* was maintained in saturation conditions with no oxygen limitation owing to the active aeration provided by the stream of bubbles.

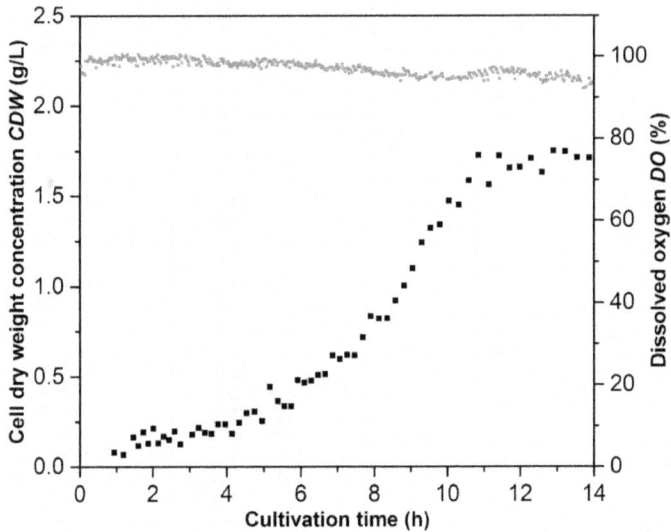

Fig. 4–11 Batch cultivation of *S. cerevisiae* in glass-based microbubble column-bioreactor (*gµBC*) with online monitoring of dissolved oxygen (*DO*) (●) and biomass concentration measured as cell dry weight (*CDW*) concentration (■) (Lladó Maldonado et al. 2018).

5 Online sensor integration of a multiphase microbioreactor

To further develop the *μBC* concept, this chapter presents a novel disposable, fully online sensor-equipped cuvette-based microbubble column-bioreactor (*cμBC*) (with a working volume of 550 μL) with high-throughput potential due to its versatility to be used in parallel and automated, as a screening platform for biotechnological research purposes. This work includes its design, fabrication (chapter 3.1.1) and mixing and oxygen transfer characteristics as well as the integration of miniaturised optical and electrochemical sensors, which allow for the real-time online monitoring of the main cultivation process parameters, including the *OD*, *DO*, *pH*, and glucose concentration. As a demonstration, example batch cultivations of *S. cerevisiae* were performed. Validation through these batch cultivations proved the long-term functionality of the sensors and reactor and established that process variable evolution could be observed over time.

5.1 Microbioreactor characterization

Among the parameters measured to characterize the functionality of the *cμBC* were the air flow rate (together with the superficial liquid velocity) and k_La. The k_La was measured in triplicate via the dynamic gassing-out method (Fig. 5–1) using the air flow rate applied during cultivation, resulting in a value of 204 1/h with an OTR_{max} of 48 mmol/(L·h). These values were comparable to the k_La values of other *MBR* or miniaturised systems described in literature (Bolic et al., 2016; Doig et al., 2005a; Doig et al., 2005b; Kheradmandnia et al., 2015; Kirk et al., 2016; Lladó Maldonado et al., 2018; Peterat et al., 2014; Weuster-Botz et al., 2001).

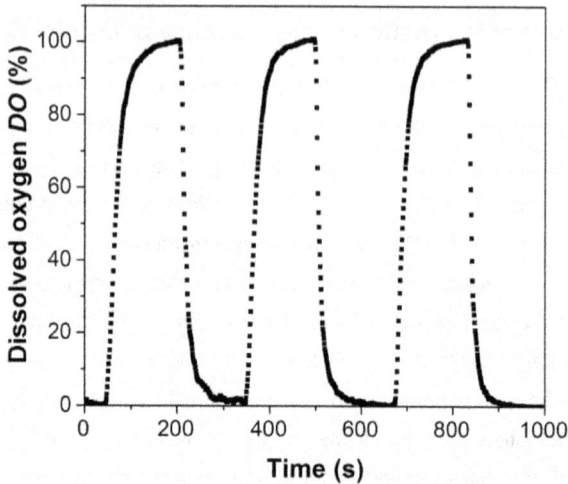

Fig. 5–1 The percentage of dissolved oxygen (*DO*) saturation changes in the cuvette-based microbubble column-bioreactor (*cµBC*) during a gassing-out experiment in triplicate for the $k_L a$ determination (Lladó Maldonado et al. 2019a).

The mixing time with 95 % homogeneity criteria θ_{95} was based on the time profile of the variation in the average colour pixel intensity in the vertical surface of the *cµBC* after the introduction of the fluorescent tracer solution, as reported in chapter 3.2.1. Fig. 5–2 shows an example sequence of images captured immediately after the injection of a fluorescent tracer pulse through a precision syringe pump. The mixing time of the *cµBC* was characterized to range from 3 to < 1 s for different air volume flow rates between 4.1 to 12.8 µL/s, which are very fast mixing times compared to other *MBRs* or miniaturised systems: from 13 down to 5.5 s in the *gµBC* (Lladó Maldonado et al., 2018); between 15 and 2 s in the *PDMS-µBC* (Peterat, 2014); mixing times of ∼ 7 s (Kirk et al., 2016); between 20 and 5 s in a miniaturised 10 mL stirred tank bioreactor (Betts et al., 2006).

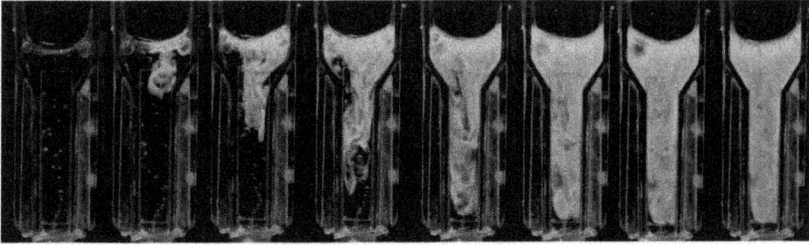

Fig. 5–2 Time-lapse image series of the *cµBC* with the air volume flow set at 6.2 µL/s and a gas superficial velocity of $3.0 \cdot 10^{-4}$ m/s after the injection of a pulse of 5 µL of the fluorescent tracer solution through a syringe pump. The images are shown with a frame rate of 3.3 fps (Lladó Maldonado et al. 2019a).

5.2 Sensor integration and calibration

The integration of optical chemical sensors for *DO* and *pH* monitoring was realized as follows. The oxygen and *pH* sensor spots were dispensed on a microscope slide as described in chapter 3.2.6 (Fig. 5–3 A). The microscope slide was then glued to a half cuvette. Subsequently, a hole was drilled into the backside of the *MBR*, and the electrochemical glucose sensor (Fig. 5–3 B), which was previously integrated into a microfluidic flow chip (Fig. 5–3 C), was attached, fixing the membrane area to the created hole.

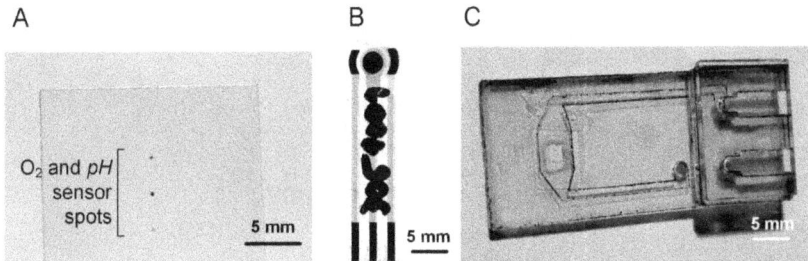

Fig. 5–3 (A) Sensing material applied (oxygen and *pH* sensor spots) onto the substrate with a microdispenser, (B) glucose sensor, and (C) two-dimensional microfluidic flow chip (Lladó Maldonado et al. 2019a).

The oxygen sensor spots were calibrated via a simple two-point calibration. For the *pH* sensor spots, a 5-point calibration was carried out with cultivation media adjusted to different *pH* values as mentioned in chapter 3.2.6.

A linear correlation between the glucose concentration and the current measured by the PalmSens potentiostat was proven for concentrations of up to 0.5 g/L in Panjan et al. (Panjan et al., 2017a). In the current work, higher concentrations of glucose could be measured due to the tuneable dynamic range offered by the adjustment of the microfluidic flow chip diffusion time and buffer flow rate. Fig. 5–4 shows the current signal with different diffusion times; the current signal value was larger when longer diffusion times were applied because more glucose molecules could diffuse. For cultivations using a glucose concentration of 20 g/L, a diffusion time of 7.5 min and a flow time of 1.5 min at 100 µL/min of *PBS* buffer were used. The glucose sensors required daily calibration because the *GOx* activity varied; however, a 2-point calibration was performed, which proved to be suitable for this application.

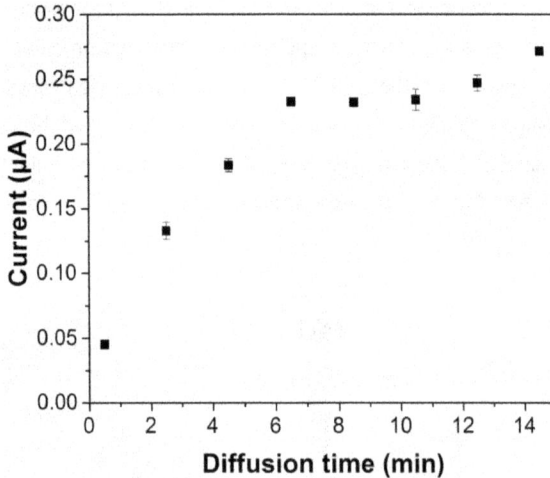

Fig. 5–4 Example of the glucose sensor signal intensity with various diffusion times (in triplicates) applied to the microfluidic flow chip buffer at a flow time of 1.5 min and flow rate of 100 µL/min when the *cµBC* was filled with cultivation medium with a glucose concentration of 5 g/L. Error bars indicate standard deviations (Lladó Maldonado et al. 2019a).

5.3 Cultivation of *Saccharomyces cerevisiae*

Batch cultivations of the model microorganism *S. cerevisiae* were performed as a test for the applicability of the fully equipped with sensors *cµBC*. *S. cerevisiae* was cultivated in the *cµBC* with an initial OD_{600} of 0.3 as described in chapter 3.2.3. With the implemented sensor system, monitoring the changes in (a) glucose concentration, (b) CDW concentration through the determination of the OD, (c) pH, and (d) DO was possible over the entire cultivation time.

In Fig. 5–5 the averaged growth of S. cerevisiae over cultivation time of two independent cultivation replicates with the respective standard deviations is presented. After an initial lag phase (cultivation time 0 - 1 h), there was a phase of exponential growth on glucose (1 – 10 h), a second lag phase due to the adaptation of S. cerevisiae to the metabolic products generated (10.5 – 12.5 h), a subsequent post-diauxic phase (12.5 – 19 h) in which the generated metabolic products (predominant metabolite ethanol) served as carbon sources and a final stationary phase (from 19 h). In the first growth phase, S. cerevisiae utilized glucose as a carbon source and produced acids and ethanol. Later, in the subsequent growth phase, when glucose was depleted (or at a low concentration), the ethanol that had accumulated served as a substrate for further growth of the cells. This diauxic behaviour was observed at 10.5 h of cultivation and, from a metabolic point of view, can be explained by the switch of the main carbon source from glucose to ethanol. During this second lag phase, the organism adapted to changes in the cultivation conditions by synthesizing new enzymes to consume the metabolites that were generated (Jones and Kompala, 1999). This behaviour has been reported by Sonnleitner and Käppeli (Sonnleitner and Käppeli, 1986) and was recently described by Bisschops et al. (Bisschops et al., 2015).

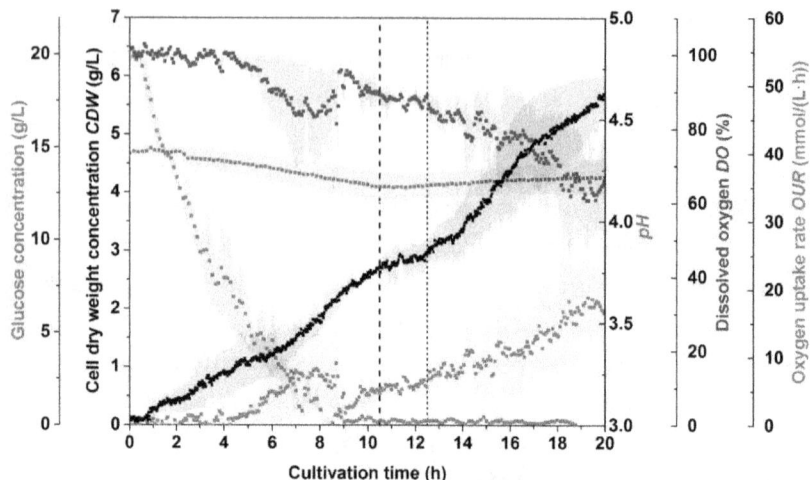

Fig. 5–5 Profile of the glucose concentration, cell dry weight (*CDW*) concentration, *pH*, dissolved oxygen (*DO*) and oxygen uptake rate (*OUR*) averaged from two independent batch cultivations of *S. cerevisiae* CCOS 538. The vertical dashed line indicates the end of the glucose-based growth and the start of the diauxic phase. The vertical pointed line indicates the end of the diauxic phase and the start of the ethanol-based growth phase. The cloud of every curve corresponds to the standard deviation of the averaged values (Lladó Maldonado et al. 2019a).

This hypothesis was confirmed by monitoring the changes in the glucose concentration and *pH*. The glucose concentration in the cultivation medium at the beginning of cultivation was 20 g/L, and the *pH* was 4.34. After 9 h of cultivation, the glucose was totally consumed, and therefore, no glucose was available for further cell growth. At 10 h, cell growth slowed down, and at 11.5 h, the *pH* stopped decreasing (*pH*= 4.17). During the first 11 h of cultivation, *S. cerevisiae* produced acidic metabolites, e.g., acetate, which accumulated in the medium, causing the *pH* to decrease. After 10 h of cultivation, because glucose, the preferred carbon source, was exhausted, the cells started utilizing the acids and other previously generated carbon sources for further growth, which explains the increase in *pH* to 4.22 at the end of cultivation. At 17 h of cultivation, cell growth slowed again, most likely due to the total consumption of generated metabolites during the first phase (e.g., ethanol and acetate), indicating the beginning of the final stationary phase. The acidification of the extracellular medium by yeast cultures has been reported previously (Castrillo et al., 1995; Sigler et al., 1981a; Sigler et al., 1981b). The proton exchange between cells and medium that is responsible for the observed variations of *pH* in biological cultures is a multicomponent process in which different mechanisms are involved. First, metabolic activity generates or consumes protons, depending on the dominant metabolic pathway of the organism at this point in cultivation. Second, the dissociation equilibrium of added substrates or products can contribute to the appearance or disappearance of protons, and third, other medium components can provide a buffering capacity (Castrillo et al., 1995). In this case, the alteration in the *pH* tendency was most likely influenced by metabolic pathway changes because they occurred simultaneously with the end of the glucose-based growth phase and the start of the ethanol-based growth phase.

During the glucose phase, the maximum specific growth rate was estimated to be μ_{max}= 0.35 1/h, reaching a final *CDW* concentration of 3 g/L, which corresponds to a cell mass yield of $Y_{X/S}$= 0.15 g_{CDW}/g_S. During the post-diauxic phase, the maximum growth rate from ethanol oxidation was μ_{max}= 0.13 1/h, which is threefold lower than that during fast growth on glucose, and a final *CDW* concentration of 5.7 g/L; the cell mass yield could not be determined due to the absence of ethanol measurements. Growth in the post-diauxic phase in which metabolism is completely respiratory is slower than that of the preceding glucose phase (Sonnleitner and Käppeli, 1986).

The value for the specific growth rate of *S. cerevisiae* on glucose (0.35 1/h) and on ethanol (0.13 1/h) are comparable to values reported in the literature for aerobic batch cultivations. For cultivations in a 1.5 L-stirred and aerated bioreactor, Bisschops et al. (Bisschops et al., 2015) reported a specific growth rate of 0.39 1/h during the glucose growth phase and 0.10 1/h during the post-diauxic phase. Beck and von Meyenburg (Beck and von Meyenburg, 1968) reported values of 0.42 1/h and 0.14 1/h, respectively, in a 3 L-stirred bioreactor. Kuhlmann et al. (Kuhlmann et al., 1984) reported specific growth rates of 0.40 1/h and 0.13 1/h, respectively, in a 2.5 L-stirred bioreactor. At the microliter-scale, Peterat et al. (Peterat et al., 2014) reported a specific growth rate on glucose of 0.37 1/h for cultivations in a 60-µL *PDMS-µBC*. Although the volumes and initial substrate concentrations were different in all these cases (10 – 30 g/L glucose), the specific growth rate values determined using the *cµBC* were roughly the same as those published from laboratory scale experiments, thus validating the applicability of the $c\mu BC$ as a suitable screening tool for aerobic submerged cultivations.

Due to the high specific oxygen demand of the cells, the DO decreased then reached a plateau at the end of the growth, after which it was observed to increase again to the saturation level (data not shown in Fig. 5–5). The DO level remained high (above 50 % air saturation) during the whole cultivation period, confirming that sufficient aeration for cultivation can be ensured by the active aeration of the $c\mu BC$ and that no oxygen limitation occurred during cultivation. Due to the small cultivation volume, pneumatic-induced aeration was a powerful and simple way to achieve sufficient oxygen supply for cultivation. The OUR was calculated from the variance in DO together with the determined k_La (estimated to be constant during the cultivation) according to eq. (2-5). The OUR during the predominantly fermentative glucose-based growth phase started at a low, non-zero value due to the relatively low-level activity of the glucose oxidative pathway. At the end of the glucose-based growth phase, the OUR increased briefly because, for a short period, the glucose oxidative pathway was preferred.

During the diauxic lag phase, the key enzymes required to catalyze the ethanol oxidative pathway were synthesized, allowing S. cerevisiae to subsequently consume the ethanol produced during the first growth phase. The OUR increased substantially during the ethanol-based growth phase, which fits with the oxidative nature of the pathway. In the final stationary phase, the OUR decreased because the metabolic activity of the cells decreased (not shown in Fig. 5–5). These observations were in good accordance to Otterstedt et al. (Otterstedt et al., 2004), which investigated the metabolism of S. cerevisiae.

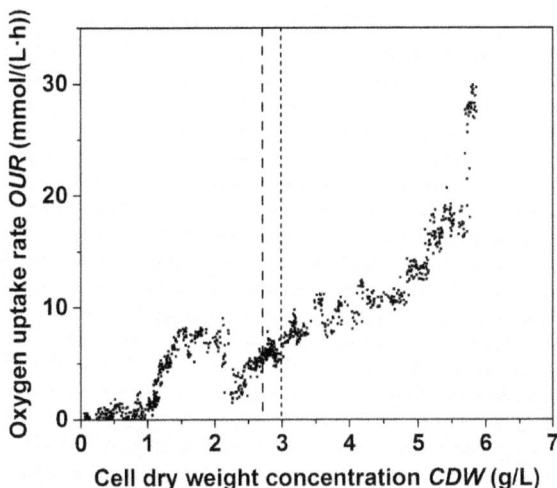

Fig. 5–6 Oxygen uptake rate (*OUR*) as a function of the cell dry weight (*CDW*) concentration averaged from two independent batch cultivations of *S. cerevisiae*. The area between vertical dashed and pointed lines indicates the start of the diauxic phase (at a *CDW* concentration of ~ 3 g/L), a short period between the end of the glucose-based growth phase and the start of the ethanol-based growth phase. The slope represents the specific oxygen uptake rate (*sOUR*), which was 2.0 - 2.5 mmol $O_2/(g_{CDW}·h)$ in the glucose-based growth phase and 2.5 - 4.5 mmol $O_2/(g_{CDW}·h)$ in the ethanol-based growth phase (Lladó Maldonado et al. 2019a).

The specific oxygen uptake rate (*sOUR*) is characteristic for each microorganism and is usually considered constant during microbial growth on a certain carbon source (Garcia-Ochoa et al., 2010). The *OUR* values increased linearly in proportion to the *CDW* concentration (Fig. 5–6). During the glucose-based growth phase, the *sOUR* was 2.0 - 2.5 mmol $O_2/(g_{CDW}·h)$, while during the ethanol-based growth phase, it was 2.5 - 4.5 mmol $O_2/(g_{CDW}·h)$. The *sOUR* in the second growth phase was higher because the metabolism of ethanol is oxidative, while during the glucose-based growth phase, the glucose fermentative pathway is preferred. The values of the *sOUR* were similar to values reported in the literature: 1.3 - 1.5 mmol $O_2/(g_{CDW}·h)$ in the glucose-based growth phase (Peterat et al., 2014) and from 2 - 12 mmol $O_2/(g_{CDW}·h)$ (von Meyenburg, 1969b) and 3 - 9 mmol $O_2/(g_{CDW}·h)$ (Kuhlmann et al., 1984) for the glucose and ethanol-based growth phases, respectively. Higher *sOUR* values have been reported under continuous cultivation conditions, with a maximum *sOUR* of up to 8.25 mmol $O_2/(g_{CDW}·h)$ achieved when the

73

dilution rate occurred at the onset of aerobic ethanol formation (von Meyenburg, 1969a; Rieger et al., 1983; Sonnleitner and Käppeli, 1986).

6 Application of microbioreactors

In chapter 5 it has been shown the development of a novel disposable, fully online sensor-equipped *cµBC* as a screening platform for biotechnological research purposes and its validation by batch cultivation of *S. cerevisiae* proving the long-term functionality of the sensors and reactor and established that variable process evolution could be observed over time (Lladó Maldonado et al., 2019a). The *cµBC* showed suitable aeration characteristics with k_La between 204 and 775 1/h. The values for the k_La varied depending on the applied airflow rate, the composition of the cultivation medium, and mixing time which ranged between 1 and 3 s. Based on these advances, the current investigations were undertaken to show the versatility of the multiphase *cµBC* by presenting two exemplary applications in different fields of bioprocesses: on one hand the function as a platform for aerobic cultivation of biological systems in batch and chemostat mode (chapter 6.1), and on the other hand its function as a research tool in biocatalysis (chapter 6.2).

The *cµBC* was equipped with miniaturised optical and electrochemical sensors that allowed the real-time online monitoring of the optical density *(OD)*, dissolved oxygen *(DO)* and glucose concentration. Depending on the application the configuration of the *cµBC* changed. Tab. 6–1 describes the configurations of the *cµBC* adopted for every application and the methodology used to monitor the bioprocess variables.

Tab. 6–1 Configurations of the cuvette-based microbubble column-bioreactor *(cµBC)* and measurement methods of the bioprocess variables.

	Batch mode	Continuous mode
Cultivation in *cµBC*	Closed system OD online: minispectrometer DO online: needle sensor Glucose online: glucose biosensor	Continuous flow OD offline: spectrometer DO online: needle sensor Glucose offline: glucose analyser
Enzymatic reaction in *µFBR*	Closed system DO online: needle sensor Glucose online: glucose biosensor	Continuous flow Retention of the immobilizates inside the *µFBR* with a membrane in the outlet DO online: needle sensor Glucose offline: HPLC measurements

6.1 Cultivation in batch and continuous mode of *Staphylococcus carnosus**

The first application of the present study was to validate the functionality of the novel *cµBC* as a tool for screening reaction kinetics in batch and chemostat. The batch cultivation of *S. carnosus* was performed in the *cµBC* as described in chapter 3.2.4 with the online monitoring of *OD*, *DO* and glucose concentration over the entire cultivation time.

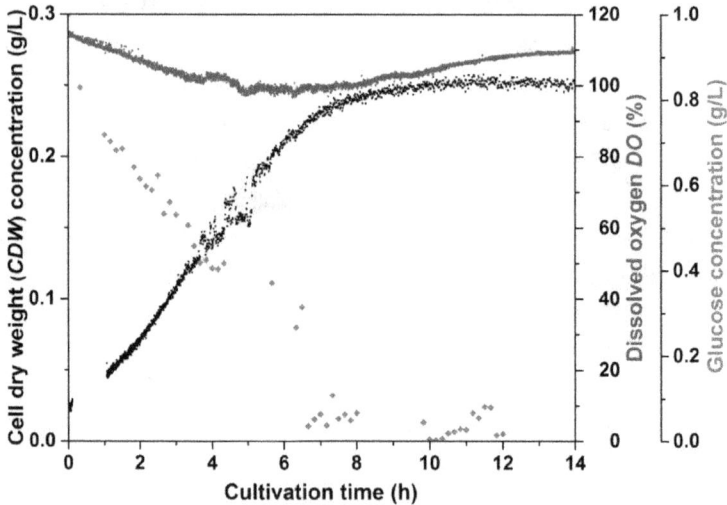

Fig. 6–1 Batch cultivation of *S. carnosus* in the cuvette-based microbubble column-bioreactor (*cµBC*) with measurement of the cell dry weight (*CDW*) concentration (■), dissolved oxygen (*DO*) (●), and glucose concentration (♦) over time (Lladó Maldonado et al. 2019b).

In Fig. 6–1 the batch cultivation *of S. carnosus* over time for one of the cultivation replicates is presented. After an initial lag phase, there was a phase of exponential growth phase with a maximum growth rate of μ= 0.39 1/h and reaching a final biomass concentration of 0.27 g/L after 8 h of cultivation, where the glucose was fully depleted from the cultivation medium. The *DO* decreased slightly during the exponential phase, and increased again during the stationary phase. The cultivation medium was almost saturated during the whole cultivation. The maximal growth rate

---

* The results were performed in the master thesis of Jana Krull in cooperation with the working group of Prof. Szita and Dr. Marques, Microfluidics Laboratory in Biochemical Engineering, University College London, London, United Kingdom (Krull, 2017).

μ_{max} values determined via batch cultivation of *S. carnosus* could be found in literature. μ_{max} was reported in Davies et al. (Davies et al., 2013) in the *MBR* "cassette" with μ_{max}= 0.60 1/h and in shaken flask μ_{max}= 0.65 1/h using the same medium of this study; and that determined by Dilsen (Dilsen et al., 2001) in a miniaturised bubble column-bioreactor with a volume of 200 - 400 mL with μ_{max}= 0.75 1/h.

To determine the values for the kinetic parameters μ_{max}, K_S, $Y_{X/S}$ and m_S continuous cultivation of *S. carnosus* was performed as described in chapter 3.2.4 at fifteen different dilution rates ranging from 0.12 to 0.80 1/h. A glucose feed concentration of c_S= 1 g/L was provided and a liquid reaction volume of V_L = 550 µL and a superficial gas velocity u_G of $2.25 \cdot 10^{-3}$ m/s were set.

The *DO* was additionally measured at some points of the cultivation. At dilution rates below 0.6 1/h where the *CDW* concentration was higher, there was a higher oxygen consumption leading to a *DO* of 60-70 %, while when the cultivation was working at higher dilution rates, the *DO* increased to 90 %, due to the reduction of the biomass concentration. Overall, the *DO* levels were more than sufficient to guarantee the optimal growth of *S. carnosus* without oxygen limitations.

Samples at each dilution rate after at least three residence periods were collected. The OD_{photo} and the glucose concentration were measured offline. OD_{photo} was converted to biomass concentration (*CDW* concentration) with the correlation described in eq. (3-11). During the continuous cultivation, the *DO* was measured at some points to prove that an adequate oxygen supply was always guaranteed. Fig. 6–2 depicted the determined stationary concentrations of cell dry weight (*CDW*) concentration, dissolved oxygen (*DO*), glucose concentration and biomass-related productivity during the continuous cultivation of *S. carnosus* at different dilution rates.

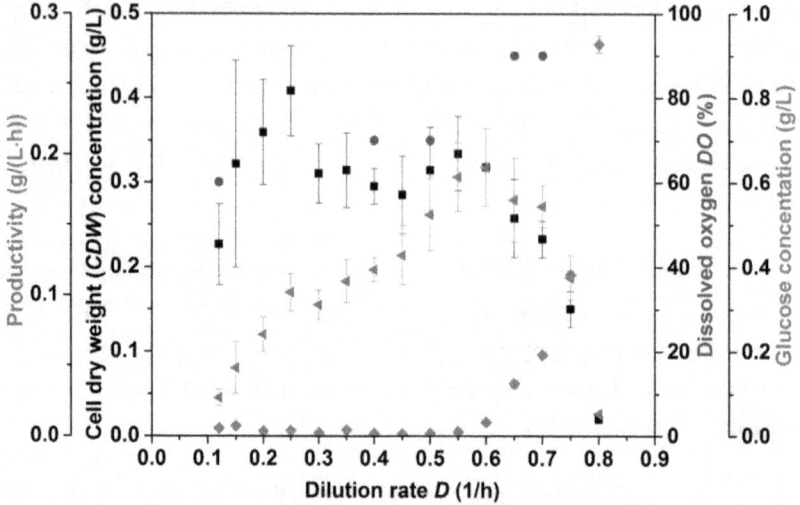

Fig. 6–2 Continuous cultivation of *S. carnosus* at different dilution rates from *D*= 0.12 to 0.80 1/h with a glucose-feed concentration of 1 g/L. Determined stationary concentrations of the cell dry weight (*CDW*) concentration (■), dissolved oxygen (*DO*) (●), glucose concentration (♦) and biomass-related productivity (◄) were calculated by eq. (2-21) (Lladó Maldonado et al. 2019b).

In the dilution rate range *D*= 0.15 - 0.55 1/h *CDW* concentration stayed approximately constant at 0.3 g/L. At D ≥ 0.6 1/h the stationary *CDW* concentrations reached lower values. The reduction of the biomass concentration was reflected in a lower glucose consumption. The biomass-related productivity was determined with eq. (2-21).

From the stationary glucose concentrations data obtained at the different dilution rates the Monod kinetic growth parameters μ_{max} and K_S were estimated with a nonlinear fitting using the Monod growth function of OriginPro 2015 (OriginLab Corporation, Northampton, Massachusetts, USA) (Fig. 6–3) with the Levenberg-Marquardt non-linear least squares algorithm for iteration and resulting in μ_{max}= 0.824 ± 0.007 1/h and K_S= 0.034 ± 0.002 g/L (R^2= 0.991). K_S describes the affinity of the organism to the substrate, with high affinities for low values. The estimated K_S= 0.034 g/L indicates a high affinity of *S. carnosus* towards glucose.

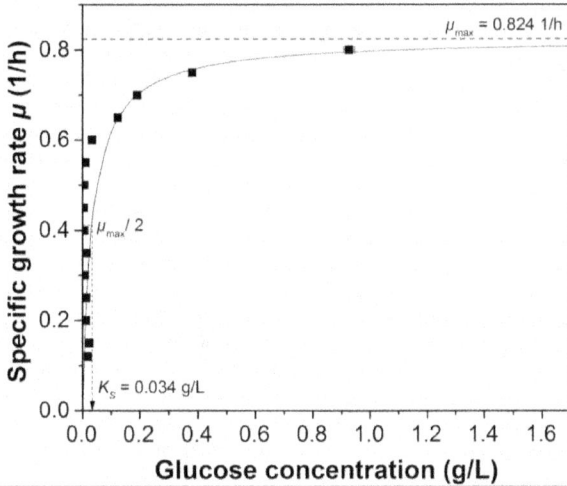

Fig. 6–3 Monod growth kinetic parameter estimation by fitting the experimental stationary glucose concentrations measured during the continuous cultivation of *S. carnosus* with a glucose-feed concentration of 1 g/L, T= 30 °C, *pH*= 6.4, at different dilution rates, which resulted in a maximum specific growth rate μ_{max}= 0.824 ± 0.007 1/h and a Monod constant K_S= 0.034 ± 0.002 g/L (R^2= 0.991) (Lladó Maldonado et al. 2019b).

In Tab. 6–2 the K_S and μ_{max} values determined via the traditional linearization methods Lineweaver-Burk, Eadie-Hofstee and Hanes-Woolf are also shown, demonstrating a good accordance between them and with the before mentioned non linear fitting method.

Tab. 6–2 The kinetic parameters maximum specific growth rate, μ_{max}, and the Monod constant, K_S, determined by linearization methods and non-linear fitting.

Method / Parameter	μ_{max} (1/h)	K_S (g/L)	R^2
Lineweaver-Burk	0.823	0.033	0.992
Eadie-Hofstee	0.823	0.034	0.989
Hanes-Woolf	0.827	0.036	0.999
Monod non linear fitting	0.824	0.034	0.991

The data of biomass and substrate concentration were used to calculate the maintenance m_S and yield coefficient $Y_{X/S}$ with eq. (2-20) by plotting the specific consumption rate q_S against D (Fig. 6–4).

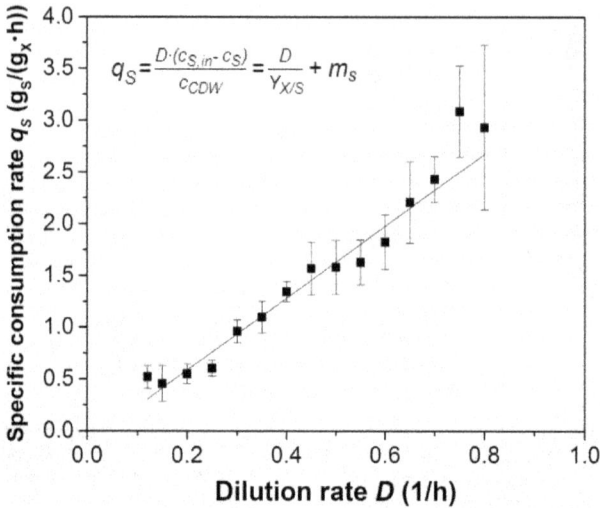

$$q_S = \frac{D \cdot (c_{S,in} - c_S)}{c_{CDW}} = \frac{D}{Y_{X/S}} + m_S$$

Fig. 6–4 The stationary specific substrate consumption rate q_S (eq. (2-20)) during the continuous cultivation ($\mu = D$) of S. carnosus at different dilution rates. By performing a linear regression analysis of the plotted data, the substrate-related biomass yield coefficient $Y_{X/S}$= 0.315 g_{CDW}/g_S (the inverse of the slope) and the maintenance coefficient m_S= 0.0035 g_S/(g_X·h) (from the intercept) were calculated (R^2= 0.983). Adapted from Lladó Maldonado et al. 2019b.

The theoretical trends for biomass and substrate concentration and productivity (solid lines) plotted in Fig. 6–5 were calculated using the previous estimated kinetic parameters (μ_{max}, K_S, $Y_{X/S}$, and m_S) together with the eq. (2-17) and (2-18). The measured experimental data (points) for the continuous cultivation of S. carnosus as a function of the dilution rate D is also depicted in Fig. 6–5.

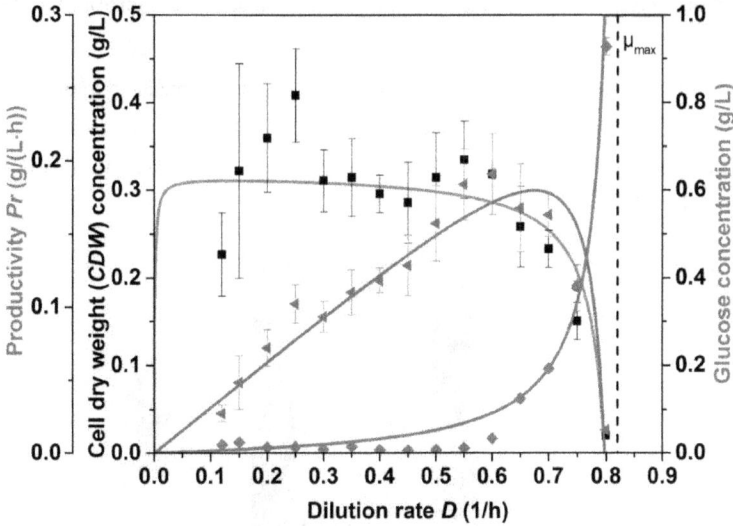

Fig. 6–5 Comparison of the trends estimated with the reaction kinetic model, using the steady state cell dry weight (*CDW*) concentration, glucose concentration (c_S) and biomass-related productivity (*Pr*) according to eq. (2-17), (2-18) and (2-21), respectively and the experimental data (c_S, ♦; c_{CDW}, ■; and *Pr*, ◄) for the continuous cultivation of *S. carnosus* as a function of the dilution rate *D*. The parameters used in the reaction kinetic model were $c_{S,in}$= 1 g$_S$/L, μ_{max}= 0.824 1/h, K_S= 0.034 g$_S$/L, $Y_{X/S}$= 0.315 g$_{CDW}$/g$_S$, and m_S= 0.0035 g$_S$/(g$_{CDW}$·h). The vertical dashed line indicates μ_{max}= 0.824 1/h (Lladó Maldonado et al. 2019b).

The dilution rate with highest biomass-related productivity $D_{Pr,max}$ is determined with eq. (2-22) with $D_{Pr,max}$= 0.675 1/h. Here the corresponding maximum cell productivity resulted in 0.18 g$_{CDW}$/(L·h). The washout value $D_{washout}$ was calculated using eq. (2-19) as $D_{washout}$= 0.797 1/h.

The experimental points fitted very well to the estimated theoretical trends of biomass, glucose and productivity as well as $D_{Pr,max}$ and $D_{washout}$ calculated from the Monod based reaction kinetic model. With this experiment the *cμBC* was proved for estimating growth kinetics rapidly and cost-effectively with continuous cultivations.

6.2 Biotransformation in batch and continuous mode in a microfluidised bed bioreactor[**]

The second application of the *cµBC* presented in this work consisted in performing a heterogeneous reaction, by introducing enzyme loaded immobilizates, fluidising them, and therefore using the *cµBC* as a microfluidised bed bioreactor (*µFBR*) as it is shown in Fig. 6-6.

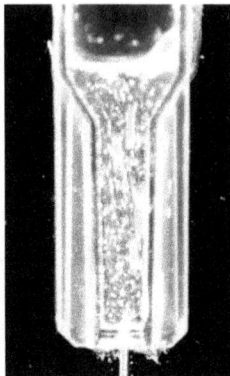

Fig. 6–6 Fluidisation in the *µFBR* of enzyme loaded immobilizates with the carrier ReliZyme 403 S.

In heterogeneous catalysis, the phase of the catalyst, which is responsible for reducing the activation energy of a reaction, is different from the phase of its reactants or substrates. Applying this concept to biocatalysis yields a system in which the enzyme is immobilized on a solid support. The advantages of working with heterogeneous catalysts are the retention and recovery of the immobilizates, the reusability of enzyme since it is no longer soluble, the better separation of enzymes from the target product, the possibility of continuous processing, and the enhanced stability. The innovative *µFBR* was validated by the study of oxygen dependent biotransformations of the model enzyme glucose oxidase (*GOx*). The insertion of immobilized GOx particles inside the *µFBR* was carried out and thus enabled the activity determination of various enzyme loadings on solid supports by the online

---

[**] The results were performed in the master thesis of R. Leopold Heydorn in cooperation with the working group of Prof. Nidetzky and Dr. Bolivar, Institute of Biotechnology and Biochemical Engineering, Graz University of Technology, Graz, Austria (Heydorn, 2017).

monitoring of the oxygen and glucose consumption. Heterogeneous biocatalysis performance with porous carrier materials is often attenuated by internal mass transfer limitations and thus insufficient oxygen supply to the internally immobilized oxygen-dependent enzymes (Bolivar et al., 2013). Besides validating the application of enzymes inside the μFBR, the long-term applicability of the GOx activity in the μFBR was investigated. It was expected as a result of the practically air saturated substrate bulk solution during the enzymatic reaction owing to the continuous air supply. This would cope with the oxygen limitation inside the porous immobilizate particles, by providing the highest possible diffusion velocity of oxygen due to the gradient between the oxygen concentrations in the bulk and inside the porous of the solid supports and this could increase the overall activity to some extent. Also an experiment in continuous mode is presented.

The data of immobilization yield, observed activity against loaded activity, and effectiveness factor for the tested supports were calculated according to Bolivar and Nidetzky (Bolivar and Nidetzky, 2012) and shown in Fig. 6–7, Fig. 6–8 and Fig. 6–9, respectively.

The immobilization yield is the percentage ratio of bound and offered protein or activity. Fig. 6–7 shows the loading capacity of the chosen enzyme carrier materials. The same weight of particles was used throughout all experiments and their diameters differed remarkably from each other. Solid supports with an increasing particle diameter led to a decreased immobilization yield. MSU-F was the smallest of the tested enzyme carrier materials and thus showed the highest amount of bound GOx due to its small size and thus high available surface area. The biggest particles used during this study have the lowest enzyme binding potential whereas ReliZyme 403 S performed better for higher enzyme loadings than Sepabeads EC-EP. This difference seems likely to be due to the different pore sizes which are bigger for ReliZyme 403 S.

83

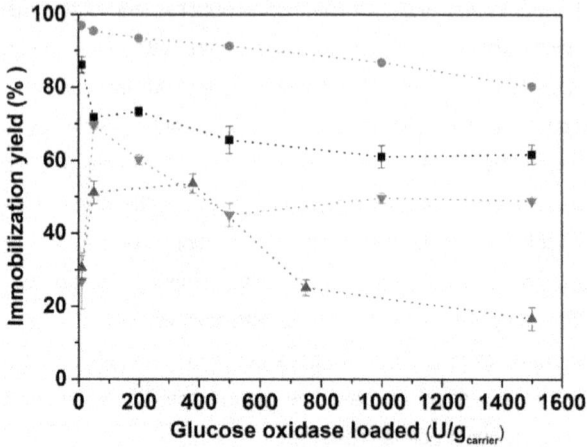

Fig. 6–7 Immobilization yield of GOx for the different supports (■ CPG 300 Å, ▼ ReliZyme 403 S, ● MSU-F, ▲ Sepabeads EC/EP).

Fig. 6–8 shows that the best resulting activities belong to the MSU–F particles which are in accordance with the high immobilization yield in Fig. 6–7 and confirms the high suitability of this material for enzyme immobilization which was already stated by Bolivar et al. (Bolivar et al., 2015).

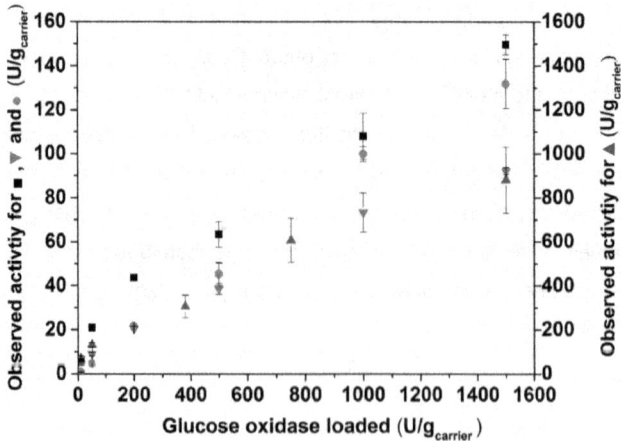

Fig. 6–8 Observed GOx activity after immobilization on different supports (■ CPG 300 Å, ▼ ReliZyme 403 S, ● MSU-F, ▲ Sepabeads EC/EP).

Bringing the final apparent enzyme carrier activities (Fig. 6–8) in relation to the previously described immobilization yield (Fig. 6–7), the effectiveness factor usually highlights the true potential of a certain material. Fig. 6–9 shows the effectiveness factor of immobilized GOx on the different supports. MSU-F is the solid support with the best properties for enzyme immobilization and high final activities since the effectiveness factor is except for the very first value in the range of 100 % or even slightly higher. In comparison, the other materials show the typical decreasing effectiveness factor for higher enzyme loadings which indicates that only a minor amount of the immobilized GOx is catalysing an enzymatic reaction.

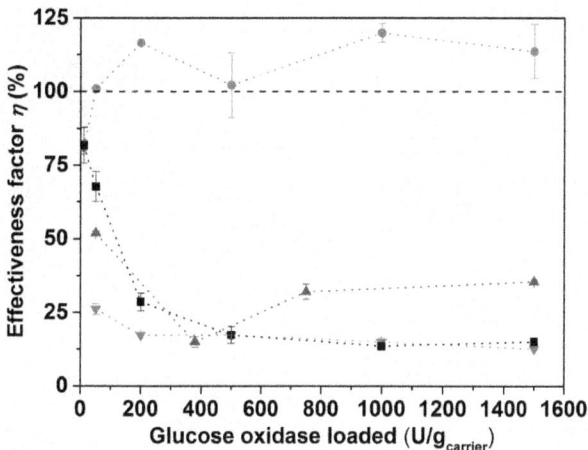

Fig. 6–9 Effectiveness factor of immobilized GOx on different supports (■ CPG 300 Å, ▼ ReliZyme 403 S, ● MSU-F, ▲ Sepabeads EC/EP).

After completing the preparation of the immobilizates and their characterisation for immobilization yield, observed activity and effectiveness factor η (chapter 3.2.5), the biocatalysts were applied and their activity checked in the μFBR. The biocatalytic activity of the enzyme loaded carriers was determined in the μFBR by using the glucose sensor and the oxygen needle sensor as described in chapter 3.2.5. Depending on the adjusted air pressure, the resulting $k_L a$ had values between 520 and 775 1/h (meaning an OTR_{max} between 123 and 183 mmol/(L·h), respectively) for a reaction volume of approximately 550 µL with 100 mM glucose and 50 mM KH_2PO_4.

In Fig. 6–10 the enzymatic activities in two different reaction systems, *μFBR* and *mSTR*, with the enzyme carriers Relizyme 403 S (Fig. 6–10 A) and MSU-F (Fig. 6–10 B) were measured and compared. The initial enzymatic activities were calculated by the measurement of the *DO* depletion through an oxygen needle sensor in both systems. The *μFBR* was equipped also with a glucose sensor, so the activity in this system was measured through the two sensors. *DO* was measured every second, while the glucose sensor was able to measure every 3 min, due to the dynamic measurement method. The enzymatic activities measured with the glucose and the oxygen sensors in the *μFBR* (ordinate) are represented against the enzymatic activities measured with the oxygen sensor in the *mSTR* (abscissa). The more similar the activities for both systems are, the closer the scattered points will be to the line *x= y*.

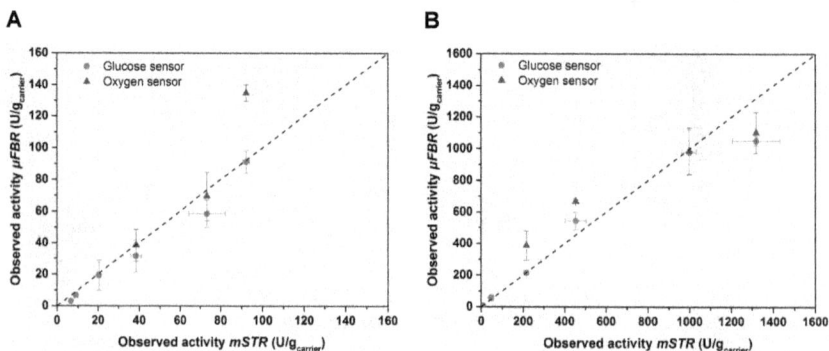

Fig. 6–10 Initial biocatalytic activities, mesured by the glucose sensor and oxygen sensor, of immobilized enzyme *GOx* on ReliZyme 403 S (A) and on MSU-F (B) carriers in the *μFBR* compared to that in the *mSTR*.

The scattered points in Fig. 6–10 are in both cases (one exception each) reasonably close to the line *x= y*, showing a good agreement between the measured activities in the *μFBR* and that measured in the *mSTR*. The experiments and measurements of the enzyme activity carried out in the *μFBR* indicate a good applicability of this reaction system for the screening of oxygen dependent enzyme immobilizates. It is also observed a good agreement between the two series of points, pointing that the enzymatic activities determined with the glucose sensor and via the measurement of *DO* gave similar results compared to those obtained in a commonly used *mSTR* with an oxygen sensor. This fact validates the glucose sensor as activity measurement method. In the case of MSU-F carrier (Fig. 6–10 B), the smallest tested particle, all enzyme loadings except for the highest with 1500 U/g were in good accordance with the values obtained in the *mSTR*.

Besides showing that the initial enzymatic activities in both reaction systems, *mSTR* and *μFBR*, were very similar, the long-term operational applicability of the enzyme carrier materials, ReliZyme 403 S and MSU-F, was compared. The depletion of substrates, oxygen and glucose in the *mSTR* and *μFBR* for ReliZyme 403 S and MSU-F is shown in Fig. 6–11 A and .6–11 B, respectively. The *DO* values are shown in the unit μmol/L since the activity is calculated from the initial slope μmol/(min·L). The *mSTR* didn't contain a glucose sensor, therefore the glucose concentration in this system was deducted from the *DO* measurements.

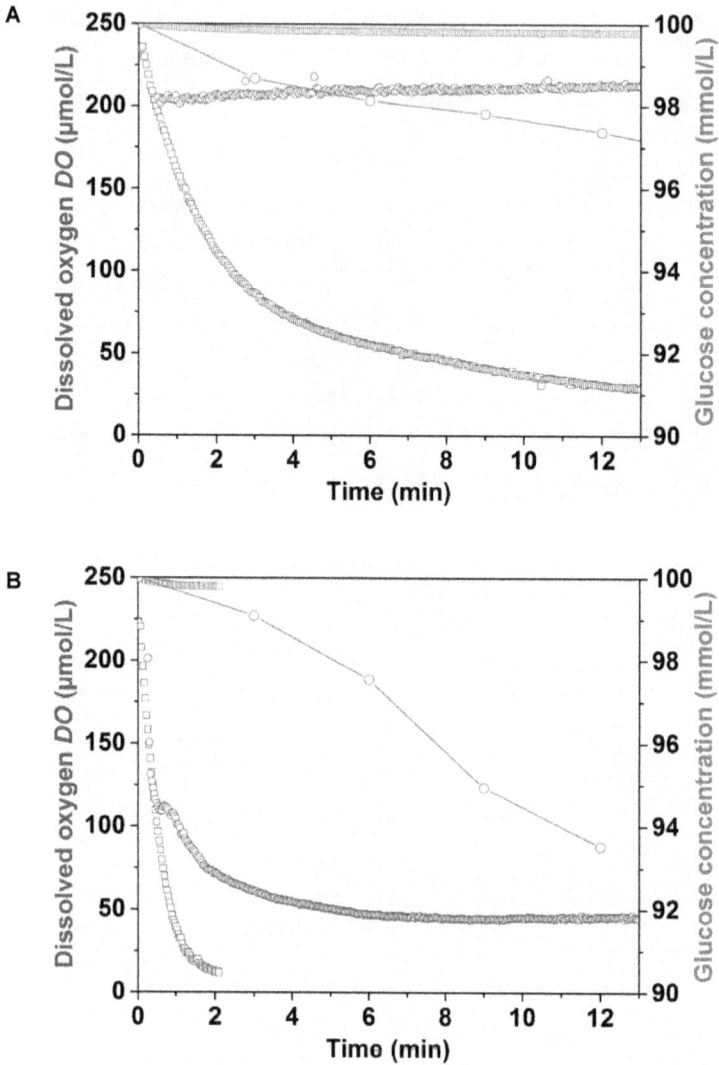

Fig. 6–11 Dissolved oxygen (*DO*) and glucose concentration over time in the bulk liquid of the *µFBR* (○) with a concentration of immobilizates of 10 mg/mL and in the *mSTR* (□) with 1.25 mg/mL during the reaction catalyzed by the immobilized enzyme on (A) ReliZyme 403 S particles (enzyme loading 500 U/g$_{carrier}$, immobilized enzyme activity 38 U/g$_{carrier}$) and on (B) MSU-F (enzyme loading 200 U/g$_{carrier}$, immobilized enzyme activity 216 U/g$_{carrier}$).

Due to the continuous oxygen supply in the μFBR, a higher carrier concentration could be introduced in the μFBR (10 mg/mL) compared to the $mSTR$ (1.25 mg/mL). In the $mSTR$ a fast initial drop of DO from the saturation concentration to DO values under 10 % was observed, while in the μFBR the DO remains approximately constant at 85 % saturation. Although the μFBR was loaded with eight times more immobilizates than the $mSTR$, the DO was maintained at a reasonable saturation level leading to a higher productivity compared to the unaerated system, which is observable in the glucose concentration after 12 min of operation, being 93.5 mmol/L and 97.3 mmol/L when working with ReliZyme and MSU-F, respectively. In the case of the $mSTR$, the glucose consumption is minimal. Here the oxygen source was just the initial oxygen saturated in the liquid. The DO concentration at 30 °C is 7.56 mg O_2/L, so 236 μM O_2, and taking into account that the reaction performed by GOx has a stoichiometric glucose : oxygen-relation of 1 : 1, the consumption of glucose could not be higher than 236 μM, by neglecting the oxygen transfer through the surface of the liquid. The improvement in the space-time yield in the μFBR was due to the aeration of the system not due to the change in the scale. The constant regeneration of oxygen allowed the long-term operational activity in the μFBR, while the $mSTR$ was just indicated for enzyme characterization studies. The reaction volume of the μFBR is a tenth of that needed in the $mSTR$, what makes it interesting for further screening studies saving reagents.

With the objective to gather and compare all carrier performances in the same figure the OUR obtained from the oxygen consumption determined in the μFBR is presented against the volumetric enzyme activity characterized in the $mSTR$ system (Fig. 6–12). The volumetric activity referred to the activity per mL of reaction volume characterized in the $mSTR$ system, while the OUR was obtained from the oxygen consumption in the μFBR. Using volumetric enzyme activity instead of enzyme loaded carrier (in $U/g_{carrier}$) is useful when comparing different carriers, because they experienced different immobilization and effectiveness yields. For every carrier (CPG 300 Å, ReliZyme 403 S MSU-F, and Sepabeads EC/EP) three points were presented, that belonged to the three different enzyme loadings that were offered. Just the data of the higher loaded carriers could be presented here, where it was possible to observe a decrease on the oxygen saturated medium, that together with the k_La data made possible to determine the OUR.

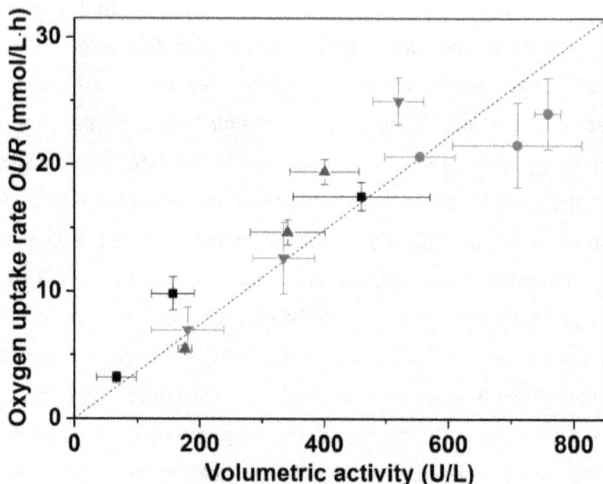

Fig. 6–12 Correlation between volumetric enzyme activity and oxygen uptake rate (OUR) derived of the characterization of the immobilizates (■ CPG 300 Å, ▼ Relizyme 403 S, ● MSU-F, ▲ Sepabeads EC/EP).

A linear correlation between OUR and volumetric activity can be observed. Since the enzyme consumes oxygen, this is reflected at the same time in the volumetric activity and in the OUR. The results shown here are helpful for comparing when using these immobilizates in other bioreactor systems, in different concentrations, different bioreactor scale or different aeration rates. Furthermore, it helps to check the minimum OTR for given volumetric enzyme activity to avoid oxygen limitations in the bulk medium. The maximum enzymatic activity that can be applied can be estimated.

For the application of the µFBR working with enzyme immobilizates in continuous mode the design was slightly modified by introducing a membrane (grade 520B, Whatman, Maidstone, United Kingdom) made of cellulose on the outlet to retain the particles inside the reaction chamber. In continuous mode it was possible to check the operational stability of the enzyme immobilizates. The µFBR was tested with immobilized GOx on Relizyme 403 S in continuous mode by running at a dilution rate of $D=$ 12 1/h (flow rate of 2 µL/s) with a 100 mM glucose solution buffered with 50 mM KH_2PO_4 (*pH 7*) [***].

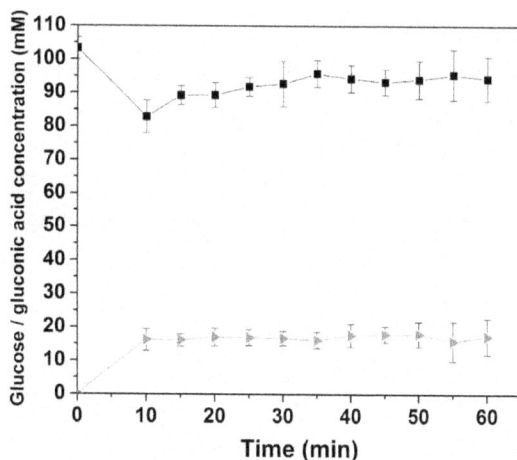

Fig. 6–13 Glucose (■) and gluconic acid (▶) concentration over time in the effluent of the *µFBR* running in continuous mode at a dilution rate of $D=$ 12 1/h (flow rate of 2 µL/s) with GOx immobilizates on Relizyme 403 S (enzyme loading 1000 $U/g_{carrier}$, immobilized enzyme activity 60 $U/g_{carrier}$). Errors bars correspond to triplicate experiments.

The continuous reaction over 60 min was feasible and evident since the glucose concentration in the effluent was lower than the substrate inlet concentration, gluconic acid concentration was measurable in the effluent, and both signals reached constant values. The gluconic acid production appeared to be very constant during the whole period. There was a slight decrease in glucose consumption in the range of 10 - 30 min that could be attributed to enzyme inactivation caused by shear stress and hydrophobic gas/ liquid interfaces which are known to occur for both soluble and externally immobilized enzymes (Betancor et al., 2005; Ghadge et al., 2003). Although the slight mismatch in the total mass balance, the enzymatic activity was proved to be stable during the whole continuous experiment and successfully retained in the *µFBR* as it is shown in Fig. 6-13.

[***] The results shown here were performed in the master thesis of Marie-Christine Mohr in cooperation with the working group of Prof. Nidetzky and Dr. Bolivar, Institute of Biotechnology and Biochemical Engineering, Graz University of Technology, Graz, Austria (Mohr, 2018).

The validation of *μFBR* in continuous processing offers several advantages when compared to batch mode. From the operational point of view, the continuous removal of media, and the possibility to adjust the substrate concentration in the inlet and the dilution rate, reduces the effect of enzyme inhibition or deactivation by substrate or product. Moreover, the retention and the recycle of the enzyme, reduces the costs and increases the productivity achieved by every g of biocatalyst.

The successful application of the *cμBC* and *μFBR* towards cultivation and enzymatic biotransformation in continuous mode highlights its promising potential for screening applications or the small scale production of e.g. active pharmaceutical ingredients or high value molecules.

7 Conclusions and outlook

The aim of the thesis was the development of *MBR* and the sensor integration for monitoring optical density (*OD*), dissolved oxygen (*DO*), *pH* and glucose as well as their validation for different biotechnological applications.

On the basis of the developed *PDMS*-based *MBR* from the *mikroPART* project, a borosilicate glass-based microbubble column-biroeactor (*gµBC*) (working volume of 60 µL, aeration occurs through a nozzle with Ø =26 µm) was designed and manufactured by wet etching and powder blasting technology (chapter 3.1.1). The investigations described the fluid dynamic and mass transfer characteristics of the *gµBC* (chapter 4), and it proved to have good oxygen transfer capacity, reaching k_La values up to 320 1/h and fast mixing times θ_{95} down to 5.5 s. The mixing performance was simulated using a simplified *CFD* model, and the tracer profile yielded a good qualitative prediction that was comparable to the experimental results, presenting a tolerable deviation of the mixing times. The development of this simplified model will provide the opportunity to analyse different reactor designs optimizing the geometry and flow conditions in further *MBR* design-manufacturing iterations, as well as acting as a starting reference for improving the model in future simulations.

The second prototype developed in this work (chapter 3.1.1) was a fully sensor-equipped *cµBC* with online sensors for *pH*, *OD*, *DO* and glucose. The design of the *MBR* was optimized from previous work, and a rapid disposable custom-made version of the reactor was built (reaction volume of 550 µL, aeration by a nozzle with a hydraulic diameter of < 100 µm). The in-house construction allowed for on-demand production, with flexibility in the reaction volume, sensor disposition, and inlet and outlet location, thus making the screening process faster and cheaper. The *cµBC* showed homogeneous mixing of the cultivation medium with θ_{95} < 1 s, with high k_La up to 775 1/h (chapter 5).

The *gµBC* and *cµBC* were validated as a suitable aerobic submerged whole-cell cultivation screening tool with batch cultivations of *Saccharomyces cerevisiae* with the real-time online monitoring of *OD* and *DO* in both reactors, and additionally *pH* and glucose in the *cµBC* (chapters 4.4 and 5.3, respectively).

Process monitoring and control is important for biotechnology applications and become challenging in small-scale reactor systems that offer limited space and sample volumes. Here, a sensing strategy for the integration of optical and biochemical sensors into the cµBC for bioprocess analysis was successfully realized. Multiple online sensing systems were implemented in a small reactor to obtain several important parameters during cultivation. The cµBC was integrated with optical chemical and electrochemical sensors (chapter 5.2). Consequently, the resulting cµBC setup allowed the online monitoring of DO, pH, glucose concentration and OD. These parameters provide valuable information on cell cultivation kinetics; moreover, they validated the usage of the cµBC as a microscreening tool because the same growth characteristics were obtained with bioreactors at the conventional litre scale. The developed cµBC demonstrated all the necessary features for autonomous, inexpensive cultivation and showed potential for online analytics and parallelisation. The long-term operational constitution of the sensors and reactor was demonstrated, as was the ability to observe process variable evolution over time. Moreover, the cµBC is amenable to modification for working in batch, fed-batch or continuous mode, which offers much potential for biotechnological research and development.

The current µBC fits with the challenges of sensing technology, control strategy and standardization for the successful implementation of microfluidic devices in bioprocessing, and therefore the cµBC can be more efficiently applied in bioprocessing. Further studies should focus on the integration of more online sensors for other substrates and products in the cµBC. The real-time, inline monitoring also would enable process control in future work, for example, pH control by adding acid or base, substrate addition or DO control by adjusting the air flow rate. The sensing technology could further be extended if the cµBC were integrated within a robotic platform with at- and offline analytical equipment to enlarge the amount of available process information. The system should be designed with a higher degree of parallelisation to fill the high throughput gap

Two exemplary applications were presented to show the versatility of the cµBC (chapter 6). The applicability of the cµBC for submerged whole-cell cultivation in chemostat mode was demonstrated by cultivating Staphylococcus carnosus (chapter 6.1) and reaching steady state concentrations of biomass and substrate at different dilution rates. Actually, for the first time the kinetic growth parameters of *S. carnosus* were estimated from data pairs of biomass and substrate concentration obtained from continuous cultivation, with $\mu_{max}=$ 0.824 1/h, $K_S=$ 0.034 g_S/L, $Y_{X/S}=$ 0.315 g_{CDW}/g_S, and $m_S=$ 0.0035 g_S/($g_{CDW}\cdot$h).

The second application of the *cµBC* was together with the addition of enzyme immobilizates, its conversion into a *µFBR* (chapter 6.2) which enabled bioconversion at the microliter-scale for oxygen-dependent enzyme immobilizates which was facilitated by the constant air supply. The fluidisation of the immobilizates was feasible inside the *µFBR*, and owing to the integration of oxygen and glucose measurement technology further enabled the determination of enzymatic activities inside of it. The use of the *µFBR* for oxygen-dependent (cell-free) biocatalysis was successfully demonstrated with the example of the model enzyme glucose oxidase immobilized on supports to convert glucose via gluconolactone and hydrogen peroxide to gluconic acid in a microfluidic bed bioreactor (*µFBR*). The initial enzymatic activities inside the *µFBR* were investigated and compared to those in a higher scaled *mSTR*. The activities resulted to be similar, proving it to be a suitable screening tool for other potential oxygen-dependent enzyme immobilizates. Moreover the enzymatic activity was proved to be stable during the continuous experiment and successfully retained.

With these applications the *cµBC* proved to be highly suitable as a submerged whole-cell cultivation- and as an oxygen dependent enzyme screening platform. Due to its continuous air supply meeting the oxygen demand of the bioprocess, its integrated online measurement technology, the possibility to work in continuous by the removal of consumed media and the addition of new media, the *cµBC* constitutes a valuable tool for analysis of the reaction kinetics of biological systems and the long-term evaluations of enzyme immobilization investigations with a rapid set up and ready to run system at low operational costs.

The prospective application of *MBRs* for enzymatic conversions is of increasing interest for reasons of higher productivity due to improved mass transfer, better process control and due to its screening tool capabilities as lower amounts of media are needed and the better reproduction of higher scaled processes.

To ensure the inclusion of microfluidic technologies in the biotechnological industry, compact equipment and/or disposable and robust *MBR*s including microsensing technology for online monitoring of process variables should be designed for practical utilization proposal. Despite the evident contribution of *MBR*s for bioprocess development, most of the applications with industrial focus have been to the medical and pharmaceutical field areas aiming to understand dynamic cells regulation and production of new drugs. By making cheaper the mass production of *MBR*s, the gap between research and industrial application is getting closer. Numbering up the *MBR*s would ensure that the product quality is consistent with that obtained with the laboratory-based optimizations, and meeting at the same time with high yield productions. With a better understanding of microorganisms and their reactions in the *µBC* as well as the reactor itself, biotechnology at microscale appears very promising.

8 References

Adler P, Frey LJ, Berger A, Bolten CJ, Hansen CE, Wittmann C. 2014. The key to acetate: metabolic fluxes of acetic acid bacteria under cocoa pulp fermentation-simulating conditions. *Appl. Environ. Microbiol.* **80**:4702–4716.

Balagaddé FK, You L, Hansen CL, Arnold FH, Quake SR. 2005. Long-term monitoring of bacteria undergoing programmed population control in a microchemostat. *Science* **309**:137–140.

Beck C, von Meyenburg HK. 1968. Enzyme pattern and aerobic growth of *Saccharomyces cerevisiae* under various degrees of glucose limitation. *J. Bacteriol.* **96**:479–486.

Betancor L, Fuentes M, Dellamora-Ortiz G, López-Gallego F, Hidalgo A, Alonso-Morales N, Mateo C, Guisán JM, Fernández-Lafuente R. 2005. Dextran aldehyde coating of glucose oxidase immobilized on magnetic nanoparticles prevents its inactivation by gas bubbles. *J. Mol. Catal. B Enzym.* **32**:97–101.

Betts JI, Doig SD, Baganz F. 2006. Characterization and application of a miniature 10 mL stirred-tank bioreactor, showing scale-down equivalence with a conventional 7 L reactor. *Biotechnol. Prog.* **22**:681–688.

Bisschops M, Vos T, Martinez-Moreno R, de la Torre Cortes P, Pronk J, Daran-Lapujade P. 2015. Oxygen availability strongly affects chronological lifespan and thermotolerance in batch cultures of *Saccharomyces cerevisiae*. *Microb. Cell* **2**:429–444.

Boccazzi P, Zhang Z, Kurosawa K, Szita N, Bhattacharya S, Jensen KF, Sinskey AJ. 2006. Differential gene expression profiles and real-time measurements of growth parameters in *Saccharomyces cerevisiae* grown in microliter-scale bioreactors equipped with internal stirring. *Biotechnol. Prog.* **22**:710–717.

Bolic A, Larsson H, Hugelier S, Eliasson Lantz A, Krühne U, Gernaey KV. 2016. A flexible well-mixed milliliter-scale reactor with high oxygen transfer rate for microbial cultivations. *Chem. Eng. J.* **303**:655–666.

Bolivar JM, Nidetzky B. 2012. Oriented and selective enzyme immobilization on functionalized silica carrier using the cationic binding module Z_{basic2}: Design of a heterogeneous D-amino acid oxidase catalyst on porous glass. *Biotechnol. Bioeng.* **109**:1490–1498.

Bolivar JM, Consolati T, Mayr T, Nidetzky B. 2013. Quantitating intraparticle O_2 gradients in solid supported enzyme immobilizates: experimental determination of their role in limiting the catalytic effectiveness of immobilized glucose oxidase. *Biotechnol. Bioeng.* **110**:2086–2095.

Bolivar JM, Nidetzky B. 2013. Smart enzyme immobilization in microstructured reactors. *Chim. Oggi - Chem. Today* **31**:50–54.

Bolivar JM, Schelch S, Mayr T, Nidetzky B. 2015. Mesoporous silica materials labeled for optical oxygen sensing and their application to development of a silica-supported oxidoreductase biocatalyst. *ACS Catal.* **5**:5984–5993.

Bolivar JM, Krämer CEM, Ungerböck B, Mayr T, Nidetzky B. 2016a. Development of a fully integrated falling film microreactor for gas-liquid-solid biotransformation with surface immobilized O_2-dependent enzyme. *Biotechnol. Bioeng.* **113**:1862–1872.

Bolivar JM, Tribulato MA, Petrasek Z, Nidetzky B. 2016b. Let the substrate flow, not the enzyme: Practical immobilization of D-amino acid oxidase in a glass microreactor for effective biocatalytic conversions. *Biotechnol. Bioeng.* **9999**:1–8.

Borisov SM, Würth C, Resch-Genger U, Klimant I. 2013. New life of ancient pigments: Application in high-performance optical sensing materials. *Anal. Chem.* **85**:9371–9377.

Bouaifi M, Hebrard G, Bastoul D, Roustan M. 2001. A comparative study of gas hold-up, bubble size, interfacial area and mass transfer coefficients in stirred gas-liquid reactors and bubble columns. *Chem. Eng. Process.* **40**:97–111.

Brás EJ, Chu V, Aires-Barros MR, Conde JP, Fernandes P. 2016. A microfluidic platform for physical entrapment of yeast cells with continuous production of invertase. *J. Chem. Technol. Biotechnol.* **92**:334–341.

Castrillo JI, De Miguel I, Ugalde UO. 1995. Proton production and consumption pathway in yeast metabolism. A chemostat culture analysis. *Yeast* **11**:1353–1365.

Davies MJ, Nesbeth DN, Szita N. 2013. Development of a microbioreactor 'cassette' for the cultivation of microorganisms in batch and chemostat mode. *Chim. Oggi - Chem. Today* **31**:46–49.

Deckwer W-D. 1992. Bubble column reactors. Ed. RW Field. *J. Chem. Technol. Biotechnol.* Chichester New York: John Wiley & Sons.

Demming S, Sommer B, Llobera A, Rasch D, Krull R, Büttgenbach S. 2011a. Disposable parallel poly(dimethylsiloxane) microbioreactor with integrated readout grid for germination screening of *Aspergillus ochraceus*. *Biomicrofluidics* **5**:14104.

Demming S, Vila-Planas J, Aliasghar Zadeh S, Edlich A, Franco-Lara E, Radespiel R, Büttgenbach S, Llobera A. 2011b. Poly(dimethylsiloxane) photonic microbioreactors based on segmented waveguides for local absorbance measurement. *Electrophoresis* **32**:431–439.

Demming S, Peterat G, Llobera A, Schmolke H, Bruns A, Kohlstedt M, Al-Halhouli A, Klages C-P, Krull R, Büttgenbach S. 2012. Vertical microbubble column-A photonic lab-on-chip for cultivation and online analysis of yeast cell cultures. *Biomicrofluidics* **6**:34106.

Dietrich N, Mayoufi N, Poncin S, Midoux N, Li HZ. 2013. Bubble formation at an orifice: A multiscale investigation. *Chem. Eng. Sci.* **92**:118–125.

van Dijken JP, Weusthuis RA, Pronk JT. 1993. Kinetics of growth and sugar consumption in yeasts. *Antonie Van Leeuwenhoek* **63**:343–352.

Dilsen S, Paul W, Herforth D, Sandgathe A, Altenbach-Rehm J, Freudl R, Wandrey C, Weuster-Botz D. 2001. Evaluation of parallel operated small-scale bubble columns for microbial process development using *Staphylococcus carnosus*. *J. Biotechnol.* **88**:77–84.

Doig SD, Diep A, Baganz F. 2005a. Characterisation of a novel miniaturised bubble column bioreactor for high throughput cell cultivation. *Biochem. Eng. J.* **23**:97–105.

Doig SD, Ortiz-Ochoa K, Ward JM, Baganz F. 2005b. Characterization of oxygen transfer in miniature and lab-scale bubble column bioreactors and comparison of microbial growth performance based on constant k_La. *Biotechnol. Prog.* **21**:1175–1182.

Edlich A, Magdanz V, Rasch D, Demming S, Aliasghar Zadeh S, Segura R, Kähler C, Radespiel R, Büttgenbach S, Franco-Lara E, Krull R. 2010. Microfluidic reactor for continuous cultivation of *Saccharomyces cerevisiae*. *Biotechnol. Prog.* **26**:1259–1270.

Ehgartner J, Sulzer P, Burger T, Kasjanow A, Bouwes D, Krühne U, Klimant I, Mayr T. 2016. Online analysis of oxygen inside silicon-glass microreactors with integrated optical sensors. *Sensors Actuators B Chem.* **228**:748–757.

Ekambara K, Dhotre MT, Joshi JB. 2005. CFD simulations of bubble column reactors: 1D, 2D and 3D approach. *Chem. Eng. Sci.* **60**:6733–6746.

Eslahpazir M, Wittmann C, Krull R. 2011. Computational Fluid Dynamics. In: Moo-Young, M, editor. *Compr. Biotechnol. Second Ed.* Elsevier, Vol. 2, pp. 1027–1038.

Ferziger JH, Peric M. 2002. Computational Methods for Fluid Dynamics 3rd ed. Springer Berlin Heidelberg 423 p.

Garcia-Ochoa F, Gomez E. 2009. Bioreactor scale-up and oxygen transfer rate in microbial processes: an overview. *Biotechnol. Adv.* **27**:153–176.

Garcia-Ochoa F, Gomez E, Santos VE, Merchuk JC. 2010. Oxygen uptake rate in microbial processes: An overview. *Biochem. Eng. J.* **49**:289–307.

Gernaey KV, Baganz F, Franco-Lara E, Kensy F, Krühne U, Luebberstedt M, Marx U, Palmqvist E, Schmid A, Schubert F, Mandenius CF. 2012. Monitoring and control of microbioreactors: An expert opinion on development needs. *Biotechnol. J.* **7**:1308–1314.

Ghadge RS, Sawant SB, Joshi JB. 2003. Enzyme deactivation in a bubble column, a stirred vessel and an inclined plane. *Chem. Eng. Sci.* **58**:5125–5134.

Grieshaber D, MacKenzie R, Vörös J, Reimhult E. 2008. Electrochemical biosensors - Sensor principles and architectures **8**:1400–1458.

Groisman A, Lobo C, Cho H, Campbell JK, Dufour YS, Stevens AM, Levchenko A. 2005. A microfluidic chemostat for experiments with bacterial and yeast cells. *Nat. Methods* **2**:685–689.

Gruber P, Marques MPC, Szita N, Mayr T. 2017a. Integration and application of optical chemical sensors in microbioreactors. *Lab Chip* **17**:2693–2712.

Gruber P, Marques MPC, Sulzer P, Wohlgemuth R, Mayr T, Baganz F, Szita N. 2017b. Real-time pH monitoring of industrially relevant enzymatic reactions in a microfluidic side-entry reactor (μSER) shows potential for pH control. *Biotechnol. J.* **12**:1600475.

Grünberger A, Paczia N, Probst C, Schendzielorz G, Eggeling L, Noack S, Wiechert W, Kohlheyer D. 2012. A disposable picolitre bioreactor for cultivation and investigation of industrially relevant bacteria on the single cell level. *Lab Chip* **12**:2060–2068.

Grünberger A, Probst C, Heyer A, Wiechert W, Frunzke J, Kohlheyer D. 2013. Microfluidic picoliter bioreactor for microbial single-cell analysis: fabrication, system setup, and operation. *J. Vis. Exp.*:50560.

Harrison DJ, Manz A, Fan Z, Luedi H, Widmer HM. 1992. Capillary electrophoresis and sample injection systems integrated on a planar glass chip. *Anal. Chem.* **64**:1926–1932.

Hatzinikolaou DG, Hansen OC, Macris BJ, Tingey A, Kekos D, Goodenough P, Stougaard P. 1996. A new glucose oxidase from *Aspergillus niger*: Characterization and regulation studies of enzyme and gene. *Appl. Microbiol. Biotechnol.* **46**:371–381.

Haverkamp V, Hessel V, Löwe H, Menges G, Warnier MJF, Rebrov E V., de Croon MHJM, Schouten JC, Liauw MA. 2006. Hydrodynamics and mixer-induced bubble formation in micro bubble columns with single and multiple-channels. *Chem. Eng. Technol.* **29**:1015–1026.

Hegab HM, Elmekawy A, Stakenborg T. 2013. Review of microfluidic microbioreactor technology for high-throughput submerged microbiological cultivation. *Biomicrofluidics* **7**:21502.

Heydorn L. 2017. Intensified bioconversion at microscale: Application of oxygen dependent enzyme immobilizates in microfluidised bed reactors. Master thesis, Institute of Biochemical Engineering, TU Braunschweig.

Illner S, Hofmann C, Löb P, Kragl U. 2014. A falling-film microreactor for enzymatic oxidation of glucose. *ChemCatChem* **6**:1748–1754.

Jakobsen HA, Lindborg H, Dorao CA. 2005. Modeling of bubble column reactors: Progress and limitations. *Ind. Eng. Chem. Res.* **44**:5107–5151.

Jokic T, Borisov SM, Saf R, Nielsen DA, Kühl M, Klimant I. 2012. Highly photostable near-infrared fluorescent pH indicators and sensors based on BF2-chelated tetraarylazadipyrromethene dyes. *Anal. Chem.* **84**:6723–6730.

Jones KD, Kompala DS. 1999. Cybernetic model of the growth dynamics of *Saccharomyces cerevisiae* in batch and continuous cultures. *J. Biotechnol.* **71**:105–131.

Kantarci N, Borak F, Ulgen KO. 2005. Bubble column reactors. *Process Biochem.* **40**:2263–2283.

Kapusta P. 2010. Absolute diffusion coefficients: Compilation of reference data for FCS calibration. Picoquant Gmbh, Berlin, Germany.

Karnik R. 2015. Microfluidic Mixing, in *Encyclopedia of Microfluidics and Nanofluidics*. Dongqing Li, editor. New York: Springer New York.

Kheradmandnia S, Hashemi-Najafabadi S, Shojaosadati SA, Mousavi SM, Malek Khosravi K. 2015. Development of parallel miniature bubble column bioreactors for fermentation process. *J. Chem. Technol. Biotechnol.* **90**:1051–1061.

Khoshmanesh K, Almansouri A, Albloushi H, Yi P, Soffe R, Kalantar-zadeh K. 2015. A multi-functional bubble-based microfluidic system. *Sci. Rep.* **5**:9942.

Kirk T V, Szita N. 2013. Oxygen transfer characteristics of miniaturized bioreactor systems. *Biotechnol. Bioeng.* **110**:1005–1019.

Kirk TV, Marques MPC, Radhakrishnan ANP, Szita N. 2016. Quantification of the oxygen uptake rate in a dissolved oxygen controlled oscillating jet-driven microbioreactor. *J. Chem. Technol. Biotechnol.* **91**:823–831.

Klein T, Schneider K, Heinzle E. 2013. A system of miniaturized stirred bioreactors for parallel continuous cultivation of yeast with online measurement of dissolved oxygen and off-gas. *Biotechnol. Bioeng.* **110**:535–42.

Kojima H, Suzuki S. 2006. Enzymatic oxidation of glucose in a pressurized bubble column. *J. Chem. Eng. Japan* **39**:1050–1053.

Kostov Y, Harms P, Randers-Eichhorn L, Rao G. 2001. Low-cost microbioreactor for high-throughput bioprocessing. *Biotechnol. Bioeng.* **72**:346–352.

Krishna R, van Baten JM. 2001. Scaling up bubble column reactors with the aid of CFD. *Chem. Eng. Res. Des.* **79**:283–309.

Krull J. 2017. Cultivation of *Staphylococcus carnosus* at micro-scale - Influence of different reactor performances. Master thesis, Institute of Biochemical Engineering, TU Braunschweig.

Krull R, Lladó Maldonado S, Lorenz T, Büttgenbach S, Demming S. 2016. Microbioreactors. In: Dietzel, A, editor. *Microsystems Pharmatechnology. Manip. fluids, Part. droplets, cells.* Cham Heidelberg New York Dordrecht London: Springer International Publishing, pp. 99–152.

Krull R, Peterat G. 2016. Analysis of reaction kinetics during chemostat cultivation of *Saccharomyces cerevisiae* using a multiphase microreactor. *Biochem. Eng. J.* **105**:220–229.

Kuhlmann W, Meyer H-D, Bellgardt KH, Schügerl K. 1984. On-line analysis of yeast growth and alcohol production. *J. Biotechnol.* **1**:171–185.

Kunze M, Lattermann C, Diederichs S, Kroutil W, Büchs J. 2014. Minireactor-based high-throughput temperature profiling for the optimization of microbial and enzymatic processes. *J. Biol. Eng.* **8**:22.

Ladner T, Grünberger A, Probst C, Kohlheyer D, Büchs J, Delvigne F. 2017. Application of mini- and micro-bioreactors for microbial bioprocesses. In: Larroche, C, Sanroman, M, Du, G, Pandey, A, editors. *Curr. Dev. Biotechnol. Bioeng.* Elsevier, pp. 433–461.

Lamping SR, Zhang H, Allen B, Ayazi Shamlou P. 2003. Design of a prototype miniature bioreactor for high throughput automated bioprocessing. *Chem. Eng. Sci.* **58**:747–758.

Lasave LC, Borisov SM, Ehgartner J, Mayr T. 2015. Quick and simple integration of optical oxygen sensors into glass-based microfluidic devices. *RSC Adv.* **5**:70808–70816.

Lattermann C, Büchs J. 2015. Microscale and miniscale fermentation and screening. *Curr. Opin. Biotechnol.* **35**:1–6.

Lee HLT, Boccazzi P, Ram RJ, Sinskey AJ. 2006. Microbioreactor arrays with integrated mixers and fluid injectors for high-throughput experimentation with pH and dissolved oxygen control. *Lab Chip* **6**:1229–1235.

Lee KS, Boccazzi P, Sinskey AJ, Ram RJ. 2011. Microfluidic chemostat and turbidostat with flow rate, oxygen, and temperature control for dynamic continuous culture. *Lab Chip* **11**:1730.

Lladó Maldonado S, Rasch D, Kasjanow A, Bouwes D, Krühne U, Krull R. 2018. Multiphase microreactors with intensification of oxygen mass transfer rate and mixing performance for bioprocess development. *Biochem. Eng. J.* **139**:57–67.

Lladó Maldonado S, Panjan P, Sun S, Rasch D, Sesay AM, Mayr T, Krull R. 2019a. A fully online sensor-equipped, disposable multiphase microbioreactor as a screening platform for biotechnological applications. *Biotechnol. Bioeng.* **116**:65–75.

Lladó Maldonado S, Krull J, Rasch D, Panjan P, Sesay A, Marques M, Szita N, Krull R. 2019b. Application of a multiphase microreactor chemostat for the determination of reaction kinetics of *Staphylococcus carnosus*. *Bioprocess Biosyst Eng.* **42**:953–961.

Löfblom J, Rosenstein R, Nguyen MT, Ståhl S, Götz F. 2017. *Staphylococcus carnosus*: from starter culture to protein engineering platform. *Appl. Microbiol. Biotechnol.* **101**:8293–8307.

Long Z, Nugent E, Javer A, Cicuta P, Sclavi B, Cosentino Lagomarsino M, Dorfman KD. 2013. Microfluidic chemostat for measuring single cell dynamics in bacteria. *Lab Chip* **13**:947–954.

Maharbiz MM, Holtz WJ, Howe RT, Keasling JD. 2003. Microbioreactor arrays with parametric control for high-throughput experimentation. *Biotechnol. Bioeng.* **85**:376–381.

Maier U, Losen M, Büchs J. 2004. Advances in understanding and modeling the gas-liquid mass transfer in shake flasks. *Biochem. Eng. J.* **17**:155–167.

Manz A, Graber N, Widmer HM. 1990. Miniaturized total chemical analysis systems: A novel concept for chemical sensing. *Sensors Actuators B Chem.* **1**:244–248.

Marbà-Ardébol AM, Emmerich J, Muthig M, Neubauer P, Junne S. 2018. Real-time monitoring of the budding index in *Saccharomyces cerevisiae* batch cultivations with in situ microscopy. *Microb. Cell Fact.* **17**:73.

Marques MPC, Cabral JMS, Fernandes P. 2010. Bioprocess scale-up: quest for the parameters to be used as criterion to move from microreactors to lab-scale. *J. Chem. Technol. Biotechnol.* **85**:1184–1198.

Marques MPC, Fernandes P, Cabral JMS, Žnidaršič-Plazl P, Plazl I. 2012. Continuous steroid biotransformations in microchannel reactors. *N. Biotechnol.* **29**:227–234.

Marques MPC, Szita N. 2017. Bioprocess microfluidics : applying microfluidic devices for bioprocessing. *Curr. Opin. Chem. Eng.* **18**:61–68.

Mateo C, Abian O, Fernandez-Lafuente R, Guisan JM. 2000. Reversible enzyme immobilization via a very strong and nondistorting ionic adsorption on support-polyethylenimine composites. *Biotechnol. Bioeng.* **68**:98–105.

Matosevic S, Szita N, Baganz F. 2011. Fundamentals and applications of immobilized microfluidic enzymatic reactors. *J. Chem. Technol. Biotechnol.* **86**:325–334.

Mauthe M, Yu W, Krut O, Krönke M, Götz F, Robenek H, Proikas-Cezanne T. 2012. WIPI-1 positive autophagosome-like vesicles entrap pathogenic *Staphylococcus aureus* for lysosomal degradation. *Int. J. Cell Biol.* **2012**:179207.

McClure DD, Norris H, Kavanagh JM, Fletcher DF, Barton GW. 2014. Validation of a computationally efficient computational fluid dynamics (CFD) model for industrial bubble column bioreactors. *Ind. Eng. Chem. Res.* **53**:14526–14543.

von Meyenburg HK. 1969a. Energetics of the budding cycle of *Saccharomyces cerevisiae* during glucose limited aerobic growth. *Arch. Mikrobiol.* **66**:289–303.

von Meyenburg K. 1969b. Katabolit-Repression und der Sprossungszyklus von *Saccharomyces cerevisiae*; PhD thesis, ETH Zürich.

Moffitt JR, Lee JB, Cluzel P. 2012. The single-cell chemostat: an agarose-based, microfluidic device for high-throughput, single-cell studies of bacteria and bacterial communities. *Lab Chip* **12**:1487–1494.

Mohr M-C. 2018. Continuous biocatalysis in a microfluidized bed reactor – application of coimmobilized glucose oxidase and catalase. Master thesis, Institute of Biochemical Engineering, TU Braunschweig.

Nacht B, Larndorfer C, Sax S, Borisov SM, Hajnsek M, Sinner F, List-Kratochvil EJW, Klimant I. 2015. Integrated catheter system for continuous glucose measurement and simultaneous insulin infusion. *Biosens. Bioelectron.* **64**:102–110.

Nahalka J, Dib I, Nidetzky B. 2008. Encapsulation of Trigonopsis variabilis D-amino acid oxidase and fast comparison of the operational stabilities of free and immobilized preparations of the enzyme. *Biotechnol. Bioeng.* **99**:251–260.

Niu H, Pan L, Su H, Wang S. 2009. Flow pattern, pressure drop, and mass transfer in a gas-liquid concurrent two-phase flow microchannel reactor. *Ind. Eng. Chem. Res.* **48**:1621–1628.

Oliveira AF, Pessoa ACSN, Bastos RG, de la Torre LG. 2016. Microfluidic tools toward industrial biotechnology. *Biotechnol. Prog.* **32**:1372–1389.

Otterstedt K, Larsson C, Bill RM, Ståhlberg A, Boles E, Hohmann S, Gustafsson L. 2004. Switching the mode of metabolism in the yeast *Saccharomyces cerevisiae*. *EMBO Rep.* **5**:532–537.

Panjan P, Ohtonen E, Tervo P, Virtanen V, Sesay AM. 2017a. Shelf life of enzymatic electrochemical sensors. *Procedia Technol.* **27**:306–308.

Panjan P, Virtanen V, Sesay AM. 2017b. Determination of stability characteristics for electrochemical biosensors via thermally accelerated ageing. *Talanta* **170**:331–336.

Panjan P, Virtanen V, Sesay AM. 2018. Towards microbioprocess control: An inexpensive 3D printed microbioreactor with integrated online real-Time glucose monitoring. *Analyst* **143**:3926–3933.

Park J, Wu J, Polymenis M, Han A. 2013. Microchemostat array with small-volume fraction replenishment for steady-state microbial culture. *Lab Chip* **13**:4217.

Peng XY, Li PCH. 2004. A three-dimensional flow control concept for single-cell experiments on a microchip. 1. Cell selection, cell retention, cell culture, cell balancing, and cell scanning. *Anal. Chem.* **76**:5273–5281.

Perozziello G, Møllenbach J, Laursen S, Di Fabrizio E, Gernaey K, Krühne U. 2012. Lab on a chip automates in vitro cell culturing. *Microelectron. Eng.* **98**:655–658.

Peterat G. 2014. Prozesstechnik und reaktionskinetische Analysen in einem mehrphasigen Mikrobioreaktorsystem. Ed. R. Krull. Göttingen: Cuvillier-Verlag. Vol. 75 103p. ibvt-Schriftenreihe, PhD thesis, TU Braunschweig.

Peterat G, Schmolke H, Lorenz T, Llobera A, Rasch D, Al-Halhouli AT, Dietzel A, Büttgenbach S, Klages C-P, Krull R. 2014. Characterization of oxygen transfer in vertical microbubble columns for aerobic biotechnological processes. *Biotechnol. Bioeng.* **111**:1809–1819.

Pfeiffer SA, Nagl S. 2015. Microfluidic platforms employing integrated fluorescent or luminescent chemical sensors: a review of methods, scope and applications. *Methods Appl. Fluoresc.* **3**:34003.

Pfeiffer T, Morley A. 2014. An evolutionary perspective on the Crabtree effect. *Front. Mol. Biosci.* **1**:1–6.

Probst C, Grünberger A, Braun N, Helfrich S, Nöh K, Wiechert W, Kohlheyer D. 2015. Rapid inoculation of single bacteria into parallel picoliter fermentation chambers. *Anal. Methods* **7**:91–98.

Puskeiler R, Kusterer A, John GT, Weuster-Botz D. 2005. Miniature bioreactors for automated high-throughput bioprocess design (HTBD): reproducibility of parallel fed-batch cultivations with *Escherichia coli*. *Biotechnol. Appl. Biochem.* **42**:227–235.

Rieger M, Käppeli O, Fiechter A. 1983. The role of limited respiration in the incomplete oxidation of glucose by *Saccharomyces cerevisiae*. *J. Gen. Microbiol.* **129**:653–661.

Rodriguez G, Weheliye W, Anderlei T, Micheletti M, Yianneskis M, Ducci A. 2013. Mixing time and kinetic energy measurements in a shaken cylindrical bioreactor. *Chem. Eng. Res. Des.* **91**:2084–2097.

Rosenstein R, Nerz C, Biswas L, Resch A, Raddatz G, Schuster SC, Götz F. 2009. Genome analysis of the meat starter culture bacterium *Staphylococcus carnosus* TM300. *Appl. Environ. Microbiol.* **75**:811–822.

Schäpper D, Alam MNHZ, Szita N, Lantz AE, Gernaey KV. 2009. Application of microbioreactors in fermentation process development: a review. *Anal. Bioanal. Chem.* **395**:679–695.

Schäpper D, Stocks SM, Szita N, Lantz AE, Gernaey KV. 2010. Development of a single-use microbioreactor for cultivation of microorganisms. *Chem. Eng. J.* **160**:891–898.

Schmolke H, Demming S, Edlich A, Magdanz V, Büttgenbach S, Franco-Lara E, Krull R, Klages C-P. 2010. Polyelectrolyte multilayer surface functionalization of poly(dimethylsiloxane) (PDMS) for reduction of yeast cell adhesion in microfluidic devices. Biomicrofluidics **4**:44113.

Shao N, Gavriilidis A, Angeli P. 2009. Flow regimes for adiabatic gas–liquid flow in microchannels. *Chem. Eng. Sci.* **64**:2749–2761.

Sigler K, Kotyk A, Knotková A, Opekarová M. 1981a. Processes involved in the creation of buffering capacity and in substrate-induced proton extrusion in the yeast *Saccharomyces cerevisiae*. *BBA - Biomembr.* **643**:583–592.

Sigler K, Knotková A, Kotyk A. 1981b. Factors governing substrate-induced generation and extrusion of protons in the yeast *Saccharomyces cerevisiae*. *Biochim. Biophys. Acta - Biomembr.* **643**:572–582.

Sobieszuk P, Aubin J, Pohorecki R. 2012. Hydrodynamics and mass transfer in gas-liquid flows in microreactors. *Chem. Eng. Technol.* **35**:1346–1358.

Sonnleitner B, Käppeli O. 1986. Growth of *Saccharomyces cerevisiae* is controlled by its limited respiratory capacity: Formulation and verification of a hypothesis. *Biotechnol. Bioeng.* **28**:927–937.

Strobl M, Rappitsch T, Borisov SM, Mayr T, Klimant I. 2015. NIR-emitting aza-BODIPY dyes – new building blocks for broad-range optical pH sensors. *Analyst* **140**:7150–7153.

Sun S, Ungerböck B, Mayr T. 2015. Imaging of oxygen in microreactors and microfluidic systems. *Methods Appl. Fluoresc.* **3**:34002.

Sun S. 2017. Applications of integrated optical sensors for pH and oxygen monitoring in micro(bio)reactor; PhD thesis, Graz University of Technology.

Szita N, Boccazzi P, Zhang Z, Boyle P, Sinskey AJ, Jensen KF. 2005. Development of a multiplexed microbioreactor system for high-throughput bioprocessing. *Lab Chip* **5**:819–826.

Thomsen MS, Nidetzky B. 2009. Coated-wall microreactor for continuous biocatalytic transformations using immobilized enzymes. *Biotechnol. J.* **4**:98–107.

Walisko J, Vernen F, Pommerehne K, Richter G, Terfehr J, Kaden D, Dähne L, Holtmann D, Krull R. 2017. Particle-based production of antibiotic rebeccamycin with Lechevalieria aerocolonigenes. *Process Biochem.* **53**:1–9.

Weuster-Botz D, Altenbach-Rehm J, Hawrylenko A. 2001. Process-engineering characterization of small-scale bubble columns for microbial process development. *Bioprocess Biosyst. Eng.* **24**:3–11.

Wilming A, Bähr C, Kamerke C, Büchs J. 2014. Fed-batch operation in special microtiter plates: a new method for screening under production conditions. *J. Ind. Microbiol. Biotechnol.* **41**:513–525.

Witt S, Wohlfahrt G, Schomburg D, Hecht HJ, Kalisz HM. 2000. Conserved arginine-516 of *Penicillium amagasakiense* glucose oxidase is essential for the efficient binding of beta-D-glucose. *Biochem. J.* **347**:553–559.

Yamamoto T, Fujii T, Nojima T. 2002. PDMS-glass hybrid microreactor array with embedded temperature control device. Application to cell-free protein synthesis. *Lab Chip* **2**:197–202.

Yu W, Götz F. 2012. Cell wall antibiotics provoke accumulation of anchored mCherry in the cross wall of *Staphylococcus aureus*. *PLoS One* **7**:e30076.

Zanzotto A, Szita N, Boccazzi P, Lessard P, Sinskey AJ, Jensen KF. 2004. Membrane-aerated microbioreactor for high-throughput bioprocessing. *Biotechnol. Bioeng.* **87**:243–254.

Zhang Z, Perozziello G, Boccazzi P, Sinskey AJ, Geschke O, Jensen KF. 2007. Microbioreactors for Bioprocess Development. *J. Assoc. Lab. Autom.* **12**:143–151.

Zhang Z, Boccazzi P, Choi H-G, Perozziello G, Sinskey AJ, Jensen KF. 2006. Microchemostat-microbial continuous culture in a polymer-based, instrumented microbioreactor. *Lab Chip* **6**:906–913.

Zhang Z, Szita N, Boccazzi P, Sinskey AJ, Jensen KF. 2005. A well-mixed, polymer-based microbioreactor with integrated optical measurements. *Biotechnol. Bioeng.* **93**:286–296.

Zucca P, Sanjust E. 2014. Inorganic materials as supports for covalent enzyme immobilization: methods and mechanisms. *Molecules* **19**:14139–14194.

Band 1   **Sunder, Matthias**: Oxidation grundwasserrelevanter Spurenverunreinigungen mit Ozon und Wasserstoffperoxid im Rohrreaktor. 1996. FIT-Verlag · Paderborn, ISBN 3-932252-00-4

Band 2   **Pack, Hubertus**: Schwermetalle in Abwasserströmen: Biosorption und Auswirkung auf eine schadstoffabbauende Bakterienkultur. 1996. FIT-Verlag · Paderborn, ISBN 3-932252-01-2

Band 3   **Brüggenthies, Antje**: Biologische Reinigung EDTA-haltiger Abwässer. 1996. FIT-Verlag · Paderborn, ISBN 3-932252-02-0

Band 4   **Liebelt, Uwe**: Anaerobe Teilstrombehandlung von Restflotten der Reaktivfärberei. 1997. FIT-Verlag · Paderborn, ISBN 3-932252-03-9

Band 5   **Mann, Volker G.**: Optimierung und Scale up eines Suspensionsreaktorverfahrens zur biologischen Reinigung feinkörniger, kontaminierter Böden. 1997. FIT-Verlag · Paderborn, ISBN 3-932252-04-7

Band 6   **Boll Marco**: Einsatz von Fuzzy-Control zur Regelung verfahrenstechnischer Prozesse. 1997. FIT-Verlag · Paderborn, ISBN 3-932252-06-3

Band 7   **Büscher, Klaus**: Bestimmung von mechanischen Beanspruchungen in Zweiphasenreaktoren. 1997. FIT-Verlag · Paderborn, ISBN 3-932252-07-1

Band 8   **Burghardt, Rudolf**: Alkalische Hydrolyse – Charakterisierung und Anwendung einer Aufschlußmethode für industrielle Belebtschlämme. 1998. FIT-Verlag · Paderborn, ISBN 3-932252-13-6

Band 9   **Hemmi, Martin**: Biologisch-chemische Behandlung von Färbereiabwässern in einem Sequencing Batch Process. 1999. FIT-Verlag · Paderborn, ISBN 3-932252-14-4

Band 10   **Dziallas, Holger**: Lokale Phasengehalte in zwei- und dreiphasig betriebenen Blasensäulenreaktoren. 2000. FIT-Verlag · Paderborn, ISBN 3-932252-15-2

Band 11   **Scheminski, Anke**: Teiloxidation von Faulschlämmen mit Ozon. 2001. FIT-Verlag · Paderborn, ISBN 3-932252-16-0

Band 12   **Mahnke, Eike Ulf**: Fluiddynamisch induzierte Partikelbeanspruchung in pneumatisch gerührten Mehrphasenreaktoren. 2002. FIT-Verlag · Paderborn, ISBN 3-932252-17-9

Band 13   **Michele, Volker**: CDF modeling and measurement of liquid flow structure and phase holdup in two- and three-phase bubble columns. 2002. FIT-Verlag · Paderborn, ISBN 3-932252-18-7

Band 14 **Wäsche, Stefan**: Einfluss der Wachstumsbedingungen auf Stoffübergang und Struktur von Biofilmsystemen. 2003. FIT-Verlag · Paderborn, ISBN 3-932252-19-5

Band 15 **Krull Rainer**: Produktionsintegrierte Behandlung industrieller Abwässer zur Schließung von Stoffkreisläuren. 2003. FIT-Verlag · Paderborn, ISBN 3-932252-20-9

Band 16 **Otto, Peter**: Entwicklung eines chemisch-biologischen Verfahrens zur Reinigung EDTA enthaltender Abwässer. 2003. FIT-Verlag · Paderborn, ISBN 3-932252-21-7

Band 17 **Horn, Harald**: Modellierung von Stoffumsatz und Stofftransport in Biofilmsystemen. 2003. FIT-Verlag · Paderborn, ISBN 3-932252-22-5

Band 18 **Mora Naranjo, Nelson**: Analyse und Modellierung anaerober Abbauprozesse in Deponien. 2004. FIT-Verlag · Paderborn, ISBN 3-932252-23-3

Band 19 **Döpkens, Eckart**: Abwasserbehandlung und Prozesswasserrecycling in der Textilindustrie. 2004. FIT-Verlag · Paderborn, ISBN 3-932252-24-1

Band 20 **Haarstrick, Andreas**: Modellierung millieugesteuerter biologischer Abbauprozesse in heterogenen problembelasteten Systemen. 2005. FIT-Verlag · Paderborn, ISBN 3-932252-27-6

Band 21 **Baaß, Anne-Christina**: Mikrobieller Abbau der Polyaminopolycarbonsäuren Propylendiamintetraacetat (PDTA) und Diethylentriaminpentaacetat (DTPA). 2004. FIT-Verlag · Paderborn, ISBN 3-932252-26-8

Band 22 **Staudt, Christian**: Entwicklung der Struktur von Biofilmen. 2006. FIT-Verlag · Paderborn, ISBN 3-932252-28-4

Band 23 **Pilz, Roman Daniel**: Partikelbeanspruchung in mehrphasig betriebenen Airlift-Reaktoren. 2006. FIT-Verlag · Paderborn, ISBN 3-932252-29-2

Band 24 **Schallenberg, Jörg**: Modellierung von zwei- und dreiphasigen Strömungen in Blasensäulenreaktoren. 2006. FIT-Verlag · Paderborn, ISBN 3-932252-30-6

Band 25 **Enß, Jan Hendrik**: Einfluss der Viskosität auf Blasensäulenströmungen. 2006. FIT-Verlag · Paderborn, ISBN 3-932252-31-4

Band 26 **Kelly, Sven**: Fluiddynamischer Einfluss auf die Morphogenese von Biopellets filamentöser Pilze. 2006. FIT-Verlag · Paderborn, ISBN 3-932252-32-2

Band 27 **Grimm, Luis Hermann**: Sporenaggregationsmodell für die submerse Kultivierung koagulativer Myzelbildner. 2006. FIT-Verlag · Paderborn, ISBN 3-932252-33-0

Band 28 **León Ohl, Andrés:** Wechselwirkungen von Stofftransport und Wachstum in Biofilsystemen. 2007. FIT-Verlag · Paderborn, ISBN 3-932252-34-9

Band 29 **Emmler, Markus:** Freisetzung von Glucoamylase in Kultivierungen mit *Aspergillus niger.* 2007. FIT-Verlag · Paderborn, ISBN 3-932252-35-7

Band 30 **Leonhäuser, Johannes:** Biotechnologische Verfahren zur Reinigung von quecksilberhaltigem Abwasser. 2007. FIT-Verlag · Paderborn, ISBN 3-932252-36-5

Band 31 **Jungebloud, Anke:** Untersuchung der Genexpression in *Aspergillus niger* mittels Echtzeit-PCR. 1996. FIT-Verlag · Paderborn, ISBN 978-3-932252-37-2

Band 32 **Hille, Andrea:** Stofftransport und Stoffumsatz in filamentösen Pilzpellets. 2008. FIT-Verlag · Paderborn, ISBN 978-3-932252-38-9

Band 33 **Fürch, Tobias:** Metabolic characterization of recombinant protein production in *Bacillus megaterium.* 2008. FIT-Verlag · Paderborn, ISBN 978-3-932252-39-6

Band 34 **Grote, Andreas Georg:** Datenbanksysteme und bioinformatische Werkzeuge zur Optimierung biotechnologischer Prozesse mit Pilzen. 2008. FIT-Verlag · Paderborn, ISBN 978-3-932252-40-120

Band 35 **Möhle, Roland Bernhard:** An Analytic-Synthetic Approach Combining Mathematical Modeling and Experiments – Towards an Understanding of Biofilm Systems. 2008. FIT-Verlag · Paderborn, ISBN 978-3-932252-41-9

Band 36 **Reichel, Thomas:** Modelle für die Beschreibung das Emissionsverhaltens von Siedlungsabfällen. 2008. FIT-Verlag · Paderborn, ISBN 978-3-932252-42-6

Band 37 **Schultheiss, Ellen:** Charakterisierung des Exopolysaccharids PS-EDIV von *Sphingomonas pituitosa.* 2008. FIT-Verlag · Paderborn, ISBN 978-3-932252-43-3

Band 38 **Dreger, Michael Andreas:** Produktion und Aufarbeitung des Exopolysaccharids PS-EDIV aus *Sphingomonas pituitosa.* 1996. FIT-Verlag · Paderborn, ISBN 978-3-932252-44-0

Band 39 **Wiebels, Cornelia:** A Novel Bubble Size Measuring Technique for High Bubble Density Flows. 2009. FIT-Verlag · Paderborn, ISBN 978-3-932252-45-7

Band 40 **Bohle, Kathrin:** Morphologie- und produktionsrelevante Gen- und Proteinexpression in submersen Kultivierungen von *Aspergillus niger.* 2009. FIT-Verlag · Paderborn, ISBN 978-3-932252-46-2

Schriftenreihe des Instituts für Bioverfahrenstechnik
der Technischen Universität Braunschweig

ibvt

Band 41   **Fallet, Claas**: Reaktionstechnische Untersuchungen der mikrobiellen Stressantwort und ihrer biotechnologischen Anwendungen. 2009. FIT-Verlag · Paderborn, ISBN 978-3-932252-47-1

Band 42   **Vetter, Andreas**: Sequential Co-simulation as Method to Couple CFD and Biological Growth in a Yeast. 2009. FIT-Verlag · Paderborn, ISBN 978-3-932252-48-8

Band 43   **Jung, Thomas**: Einsatz chemischer Oxidationsverfahren zur Behandlung industrieller Abwässer. 2010. FIT-Verlag· Paderborn, ISBN 978-3-932252-49-5

Band 45   **Herrmann, Tim**: Transport von Proteinen in Partikeln der Hydrophoben Interaktions Chromatographie. 2010. FIT-Verlag · Paderborn, ISBN 978-3-932252-51-8

Band 46   **Becker, Judith**: Systems Metabolic Engineering of *Corynebacterium glutamicum* towards improved Lysine Prodction. 2010. Cuvillier-Verlag · Göttingen, ISBN 978-3-86955-426-6

Band 47   **Melzer, Guido**: Metabolic Network Analysis of the Cell Factory *Aspergillus niger*. 2010. Cuvillier-Verlag · Göttingen, ISBN 978-3-86955-456-3

Band 48   **Bolten J., Christoph**: Bio-based Production of L-Methionine in *Corynebacterium glutamicum*. 2010. Cuvillier-Verlag · Göttingen, ISBN 978-3-86955-486-0

Band 49   **Lüders, Svenja**: Prozess- und Proteomanalyse gestresster Mikroorganismen. 2010. Cuvillier-Verlag · Göttingen, ISBN 978-3-86955-435-8

Band 50   **Wittmann, Christoph**: Entwicklung und Einsatz neuer Tools zur metabolischen Netzwerkanalyse des industriellen Aminosäure-Produzenten *Corynebacterium glutamicum*. 2010. Cuvillier-Verlag · Göttingen, ISBN 978-3-86955-445-7

Band 51   **Edlich, Astrid**: Entwicklung eines Mikroreaktors als Screening-Instrument für biologische Prozesse. 2010. Cuvillier-Verlag · Göttingen, ISBN 978-3-86955-470-9

Band 52   **Hage, Kerstin**: Bioprozessoptimierung und Metabolomanalyse zur Proteinproduktion in *Bacillus licheniformis*. 2010. Cuvillier-Verlag · Göttingen, ISBN 978-3-86955-578-2

Band 53   **Kiep, Katina Andrea**: Einfluss von Kultivierungsparametern auf die Morphologie und Produktbildung von *Aspergillus niger*. 2010. Cuvillier-Verlag · Göttingen, ISBN 978-3-86955-632-1

Band 54 **Fischer, Nicole**: Experimental investigations on the influence of physico-chemical parameters on anaerobic degradation in MBT residual waste. 2011. Cuvillier-Verlag · Göttingen, ISBN 978-3-86955-679-6

Band 55 **Schädel, Friederike**: Stressantwort von Mikroorganismen. 2011. Cuvillier-Verlag · Göttingen, ISBN 978-3-86955-746-5

Band 56 **Wichter, Johannes**: Untersuchung der L-Cystein-Biosynthese in *Escherichia coli* mit Techniken der Metabolom- und ^{13}C-Stoffflussanalyse. 2011. Cuvillier-Verlag · Göttingen, ISBN 978-3-86955-750-2

Band 57 **Knappik, Irena Isabell**: Charakterisierung der biologischen und chemischen Reaktionsprozesse in Siedlungsabfällen. 2011. Cuvillier-Verlag · Göttingen, ISBN 978-3-86955-760-1

Band 58 **Driouch, Habib**: Systems biotechnology of recombinant protein production in *Aspergillus niger*. 2011. Cuvillier-Verlag Göttingen, ISBN 978-3-86955-808-0

Band 59 **Gehder, Matthias:** Development and Validation of Indicators for the Production and Quality of Seed Cultures. 2011. Cuvillier-Verlag Göttingen, ISBN 978-3-86955-847-9

Band 60 **Sommer, Becky:** Methodenentwicklung zur Charakterisierung sporenbildender Pilz-Seedingkulturen. 2011. Cuvillier-Verlag Göttingen, ISBN 978-3-86955-851-6

Band 61 **Dohnt, Katrin:** Charakterisierung von *Pseudomonas aeruginosa*-Biofilmen in einem *in vitro*-Harnwegskathetersystem. 2011. Cuvillier-Verlag Göttingen, ISBN 978-3-86955-852-3

Band 62 **Greis, Tillman:** Meddling the risk of chlorinated hydrocarbons in urban groundwater. 2011. Cuvillier-Verlag Göttingen, ISBN 978-3-86955-970-4

Band 63 **David, Florian:** Holistic bioprocess engineering of antibody fragment secreting *Bacillus megaterium*. 2012. Cuvillier-Verlag Göttingen, ISBN 978-3-95404-115-2

Band 64 **Palme, Wiebke:** Taxonomische Einordnung des Polyaminopolycarbonsäure-abbauenden Stammes BNC1 und Untersuchungen zum Abbau von 1,3 Propylendiamintetraacetat. 2012. Cuvillier-Verlag Göttingen, ISBN 978-3-95404-158-9

Band 65 **Lin, Pey-Jin:** Effect of fluid dynamics on pellet morphology and product formation of *Aspergillus niger*. 2012. Cuvillier-Verlag Göttingen, ISBN 978-3-95404-181-7

Band 66   **Kind, Stefanie:** Synthetic Metabolic Engineering of *Corynebacterium glu-tamicum* for Bio-based Production of 1,5-Diaminopentane. 2012. Cuvillier-Verlag Göttingen, ISBN 978-3-95404-264-7

Band 67   **Wilk, Franziska:** Charakterisierung der Stoffströme vorbehandelter Sied-lungsabfälle in Deponiebioreaktoren. 2012. Cuvillier-Verlag Göttingen, ISBN 978-3-95404-281-4

Band 68   **Korneli, Claudia:** Target-oriented Bioprocess Optimization for Recombi-nant Protein Production in *Bacillus megaterium*. 2012. Cuvillier-Verlag Göttingen, ISBN 978-3-95404-289-0

Band 69   **Eslahpazir, Manely:** Numerical Characterization of Mechanical Stress and Flow Patterns in Stirred Tank Bioreactors. 2013. Cuvillier-Verlag Göttingen, ISBN 978-3-95404-449-8

Band 70   **Wucherpfennig, T.:** Cellular Morphology – A novel Process Parameter for the Cultivation of Eukaryotic Cells. 2013. Cuvillier-Verlag Göttingen, ISBN 978-3-95404-456-6

Band 71   **Buschke, Nele:** Bio-Nylon Monomers from Renewables using *Coryne-bacterium glutamicum*. 2013. Cuvillier-Verlag Göttingen, ISBN 978-3-95404-457-3

Band 72   **Bergmann, Sven:** Ectoine production by halotolerant microorganisms. 2013. Cuvillier-Verlag Göttingen, ISBN 978-3-95404-556-3

Band 73   **Hellriegel, Jan:** Engineering a Biofilm – Imitating Physico-Chemical Prop-erties to improve Mechanical Characterization. 2014. Cuvillier-Verlag Göt-tingen, ISBN 978-3-95404-753-6

Band 74   **Berger, Antje:** Metabolische Netzwerkanalyse uropathogener *Pseudo-monas aeruginosa*-Isolate. 2014. Cuvillier-Verlag Göttingen, ISBN 978-3-95404-762-8

Band 75   **Peterat, Gena:** Prozesstechnik und rekationskinetische Analysen in einem mehrphasigen Mikrobioreaktorsystem. 2014. Cuvillier-Verlag Göttingen, ISBN 978-3-95404 887-0

Band 76   **Godard, Thibault:** Systems biology of stress in *Bacillus megaterium* and its potential applications. 2016. Cuvillier-Verlag Göttingen, ISBN 978-3-7369-9336-5

Band 77   **Hönnscheidt, Christoph:** Entwicklung kolloiddisperser Wirkstoffformulie-rungen auf Basis von Biopolymeren. 2016. Cuvillier-Verlag Göttingen, ISBN 978-3-7369-9260-3

Band 78 **Walisko, Jana:** Morphologiebeeinflussung von *Lechevalieria aerocolonigenes* und heterologe Produktion von Rebeccamycin. 2017. Cuvillier-Verlag Göttingen, ISBN 978-3-7369-9502-4

Band 79 **Gädke, Johannes:** *In situ*-downstream processing of recombinant histidine-tagged proteins from cultivations of *Bacillus megaterium*. 2017. Cuvillier-Verlag Göttingen, ISBN 978-3-7369-9551-2

Band 80 **Lakowitz, Antonia:** Skalenübergreifende Produktion und Sekretion rekom-binanter Proteine mit Stämmen der Gattung *Bacillus*. 2017. Cuvillier-Verlag Göttingen, ISBN 978-3-7369-9576-5